REMEDIAL MATHEMATICS

REMEDIAL MATHEMATICS

BY

AMELIA ATKINSON | IVY JACKSON

KNOWLEDGE BAKERS

ISBN: 978-81-9759-577-6 (HB)

Published by : **Knowledge Bakers**
Pune, India
Email: info@knowledgebakers.com

PREFACE

Mathematics serves as a foundational subject that underpins various aspects of academic and professional pursuits, yet many individuals encounter challenges in mastering its concepts and applications. Remedial mathematics education plays a crucial role in addressing these challenges by providing targeted instruction and support to students who require additional assistance in developing their mathematical skills and confidence. As educators and learners alike navigate the complexities of mathematical understanding, it becomes imperative to cultivate effective strategies and resources for remediation that meet the diverse needs of students across different learning contexts. As we embark on this exploration of remedial mathematics education, we delve into a realm where pedagogy, psychology, and technology intersect, uncovering the principles, methodologies, and innovations that promote mathematical proficiency and success for all learners.

This book, titled "Remedial Mathematics," is crafted to provide a comprehensive resource for educators, students, and policymakers involved in remedial mathematics education. Through collaboration with leading experts, practitioners, and researchers in the field of mathematics education, we aim to offer readers a multidimensional and insightful journey through the intricacies of remedial mathematics instruction.

As editors of this volume, we have curated a collection of chapters that cover a wide range of topics in remedial mathematics education, including diagnostic assessment, instructional strategies, curriculum design, technology integration, differentiated instruction, and socio-emotional support for learners. Each chapter is designed to provide theoretical insights, practical methodologies, case studies, and examples to empower educators to effectively address the needs of students who require remediation in mathematics.

Furthermore, this book seeks to highlight the importance of equity, diversity, and inclusion in remedial mathematics education. By examining the barriers to mathematical learning faced by underserved and marginalized populations, we aim to inspire readers to adopt culturally responsive approaches and inclusive practices that promote mathematical success for all learners.

As we embark on this enlightening journey through the world of remedial mathematics education, we extend our gratitude to the contributors whose expertise and dedication have enriched this book. We hope that "Remedial Mathematics" serves as a valuable resource for educators, students, policymakers, and researchers alike, empowering them to cultivate mathematical proficiency, confidence, and enjoyment for learners of all backgrounds and abilities.

Amelia Atkinson | Ivy Jackson

CONTENTS

1

Introduction

Mathematics is the science that deals with the logic of shape, quantity and arrangement. Math is all around us, in everything we do. It is the building block for everything in our daily lives, including mobile devices, computers, software, architecture (ancient and modern), art, money, engineering and even sports.

Since the beginning of recorded history, mathematical discovery has been at the forefront of every civilized society, and math has been used by even the most primitive and earliest cultures. The need for math arose because of the increasingly complex demands from societies around the world, which required more advanced mathematical solutions, as outlined by mathematician Raymond L. Wilder in his book "Evolution of Mathematical Concepts(opens in new tab)" (Dover Publications, 2013).

The more complex a society, the more complex the mathematical needs. Primitive tribes needed little more than the ability to count, but also used math to calculate the position of the sun and the physics of hunting. "All the records — anthropological and historical — show that counting and, ultimately, numeral systems as a device for counting form the inception of the mathematical element in all cultures," Wilder wrote in 1968.

Mathematics is an area of knowledge that includes such topics as numbers (arithmetic, number theory), formulas and related structures (algebra), shapes and the spaces in which they are contained (geometry), and quantities and their changes (calculus and analysis).

Most mathematical activity involves discovering and proving properties of abstract objects by pure reasoning. These objects are either abstractions from nature, such as natural numbers or lines, or — in modern mathematics — entities that are stipulated with certain properties, called axioms. A proof consists of a succession of applications of some deductive rules to already known results, including previously proved theorems, axioms and (in case of abstraction from nature) some basic properties that are considered as true starting points of the theory under consideration. The result of a proof is called a theorem.

Mathematics is widely used in science for modeling phenomena. This enables the extraction of quantitative predictions from experimental laws. For example, the movement of planets can be accurately predicted using Newton's law of gravitation combined with mathematical computation. The independence of mathematical truth from any experimentation implies that the accuracy of such predictions depends only on the adequacy of the model for describing the reality. Inaccurate predictions imply the need for improving or changing mathematical models, not that mathematics is wrong in the models themselves. For example, the perihelion precession of Mercury cannot be explained by Newton's law of gravitation but is accurately explained by Einstein's general relativity. This experimental validation of Einstein's theory shows that Newton's law of gravitation is only an approximation, though accurate in everyday application.

Mathematics is essential in many fields, including natural sciences, engineering, medicine, finance, computer science and social sciences. Some areas of mathematics, such as statistics and game theory, are developed in close correlation with their applications and are often grouped under applied mathematics. Other mathematical areas are developed independently from any application (and are therefore called pure mathematics), but practical applications are often discovered later. A fitting example is the problem of integer factorization, which goes back to Euclid, but which had no practical application before its use in the RSA cryptosystem (for the security of computer networks).

In the history of mathematics, the concept of a proof and its associated mathematical rigour first appeared in Greek mathematics, most notably in Euclid's Elements. Mathematics developed at a relatively slow pace until the Renaissance, when algebra and infinitesimal calculus were added to arithmetic and geometry as main areas of mathematics. Since then, the interaction between mathematical innovations and scientific discoveries has led to a rapid increase in the development of mathematics. At the end of the 19th century, the foundational crisis of mathematics led to the systematization of the axiomatic method. This, in turn, gave rise to a dramatic increase in the number of mathematics areas and their fields of applications. An example of this is the Mathematics Subject Classification, which lists more than sixty first-level areas of mathematics.

Mathematics, the science of structure, order, and relation that has evolved from elemental practices of counting, measuring, and describing the shapes of objects. It deals with logical reasoning and quantitative calculation, and its development has involved an increasing degree of idealization and abstraction of its subject matter. Since the 17th century, mathematics has been an indispensable adjunct to the physical sciences and technology, and in more recent times it has assumed a similar role in the quantitative aspects of the life sciences.

In many cultures—under the stimulus of the needs of practical pursuits, such as commerce and agriculture—mathematics has developed far beyond basic counting. This growth has been greatest in societies complex enough to sustain these activities and to provide leisure for contemplation and the opportunity to build on the achievements of earlier mathematicians.

All mathematical systems (for example, Euclidean geometry) are combinations of sets of axioms and of theorems that can be logically deduced from the axioms. Inquiries into the logical and philosophical basis of mathematics reduce to questions of whether the axioms of a given system ensure its completeness and its consistency. For full treatment of this aspect, see mathematics, foundations of.

This article offers a history of mathematics from ancient times to the present. As a consequence of the exponential growth of science, most mathematics has developed since the 15th century CE, and it is a historical fact that, from the 15th century to the late 20th century, new developments in mathematics were largely concentrated in Europe and North America. For these reasons, the bulk of this article is devoted to European developments since 1500.

This does not mean, however, that developments elsewhere have been unimportant. Indeed, to understand the history of mathematics in Europe, it is necessary to know its history at least in ancient Mesopotamia and Egypt, in ancient Greece, and in Islamic civilization from the 9th to the 15th century. The way in which these civilizations influenced one another and the important direct contributions Greece and Islam made to later developments are discussed in the first parts of this article.

India's contributions to the development of contemporary mathematics were made through the considerable influence of Indian achievements on Islamic mathematics during its formative years. A separate article, South Asian mathematics, focuses on the early history of mathematics in the Indian subcontinent and the development there of the modern decimal place-value numeral system. The article East Asian mathematics covers the mostly independent development of mathematics in China, Japan, Korea, and Vietnam.

The substantive branches of mathematics are treated in several articles. See algebra; analysis; arithmetic; combinatorics; game theory; geometry; number theory; numerical analysis; optimization; probability theory; set theory; statistics; trigonometry.

HISTORY

The area of study known as the history of mathematics is primarily an investigation into the origin of discoveries in mathematics and, to a lesser extent, an investigation into the mathematical methods and notation of the past. Before the modern age and the worldwide spread of knowledge, written examples of new mathematical developments have come to light only in a few locales. From 3000 BC the Mesopotamian states of Sumer, Akkad and Assyria, together with Ancient Egypt and Ebla began using arithmetic, algebra and geometry for purposes of taxation, commerce, trade and also in the field of astronomy and to formulate calendars and record time.

The most ancient mathematical texts available are from Mesopotamia and Egypt - Plimpton 322 (Babylonian c. 1900 BC),the Rhind Mathematical Papyrus (Egyptian c. 2000–1800 BC) and the Moscow Mathematical Papyrus (Egyptian c. 1890 BC). All of these texts mention the so-called Pythagorean triples and so, by inference, the Pythagorean theorem, seems to be the most ancient and widespread mathematical development after basic arithmetic and geometry.

The study of mathematics as a "demonstrative discipline" begins in the 6th century BC with the Pythagoreans, who coined the term "mathematics" from the ancient Greek ìÜèçìá (mathema), meaning "subject of instruction". Greek mathematicsgreatly refined the methods (especially through the introduction of deductive reasoning and mathematical rigor in proofs) and expanded the subject matter of mathematics. Although they made virtually no contributions to theoretical mathematics, the ancient Romans used applied mathematics in surveying, structural engineering, mechanical engineering, bookkeeping, creation of lunar and solar calendars, and even arts and crafts. Chinese mathematics made early contributions, including a place value system and the first use of negative numbers. The Hindu–Arabic numeral systemand the rules for the use of its operations, in use throughout the world today evolved over the course of the first millennium AD in India and were transmitted to the Western world via Islamic mathematics through the work of Mu%ammad ibn Mûsâ al-Khwârizmî. Islamic mathematics, in turn, developed and expanded the mathematics known to these civilizations.Contemporaneous with but independent of these traditions were the mathematics developed by the Maya civilization of Mexico and Central America, where the concept of zero was given a standard symbol in Maya numerals.

Many Greek and Arabic texts on mathematics were translated into Latin from the 12th century onward, leading to further development of mathematics in Medieval Europe. From ancient times through the Middle

Ages, periods of mathematical discovery were often followed by centuries of stagnation. Beginning in Renaissance Italy in the 15th century, new mathematical developments, interacting with new scientific discoveries, were made at an increasing pace that continues through the present day. This includes the groundbreaking work of both Isaac Newton and Gottfried Wilhelm Leibniz in the development of infinitesimal calculus during the course of the 17th century. At the end of the 19th century the International Congress of Mathematicians was founded and continues to spearhead advances in the field.

Prehistoric

The origins of mathematical thought lie in the concepts of number, magnitude, and form. Modern studies of animal cognition have shown that these concepts are not unique to humans. Such concepts would have been part of everyday life in hunter-gatherer societies. The idea of the "number" concept evolving gradually over time is supported by the existence of languages which preserve the distinction between "one", "two", and "many", but not of numbers larger than two.

Prehistoric artifacts discovered in Africa, dated 20,000 years old or more suggest early attempts to quantify time. The Ishango bone, found near the headwaters of the Nile river (northeastern Congo), may be more than 20,000 years old and consists of a series of marks carved in three columns running the length of the bone. Common interpretations are that the Ishango bone shows either a tally of the earliest known demonstration of sequences of prime numbers or a six-month lunar calendar. Peter Rudman argues that the development of the concept of prime numbers could only have come about after the concept of division, which he dates to after 10,000 BC, with prime numbers probably not being understood until about 500 BC. He also writes that "no attempt has been made to explain why a tally of something should exhibit multiples of two, prime numbers between 10 and 20, and some numbers that are almost multiples of 10." The Ishango bone, according to scholar Alexander Marshack, may have influenced the later development of mathematics in Egypt as, like some entries on the Ishango bone, Egyptian arithmetic also made use of multiplication by 2; this however, is disputed.

Predynastic Egyptians of the 5th millennium BC pictorially represented geometric designs. It has been claimed that megalithic monuments in England and Scotland, dating from the 3rd millennium BC, incorporate geometric ideas such as circles, ellipses, and Pythagorean triples in their design. All of the above are disputed however, and the currently oldest undisputed mathematical documents are from Babylonian and dynastic Egyptian sources.

Babylonian

Babylonian mathematics refers to any mathematics of the peoples of Mesopotamia (modern Iraq) from the days of the early Sumerians through the Hellenistic period almost to the dawn of Christianity. The majority of Babylonian mathematical work comes from two widely separated periods: The first few hundred years of the second millennium BC (Old Babylonian period), and the last few centuries of the first millennium BC (Seleucid period). It is named Babylonian mathematics due to the central role of Babylon as a place of study. Later under the Arab Empire, Mesopotamia, especially Baghdad, once again became an important center of study for Islamic mathematics.

Geometry problem on a clay tablet belonging to a school for scribes; Susa, first half of the 2nd millennium BCE

In contrast to the sparsity of sources in Egyptian mathematics, our knowledge of Babylonian mathematics is derived from more than 400 clay tablets unearthed since the 1850s. Written in Cuneiform script, tablets were inscribed whilst the clay was moist, and baked hard in an oven or by the heat of the sun. Some of these appear to be graded homework.

The earliest evidence of written mathematics dates back to the ancient Sumerians, who built the earliest civilization in Mesopotamia. They developed a complex system of metrology from 3000 BC. From around 2500 BC onwards, the Sumerians wrote multiplication tables on clay tablets and dealt with geometrical exercises and division problems. The earliest traces of the Babylonian numerals also date back to this period.

The Babylonian mathematical tablet Plimpton 322, dated to 1800 BC.

Babylonian mathematics were written using a sexagesimal (base-60) numeral system. From this derives the modern day usage of 60 seconds in a minute, 60 minutes in an hour, and 360 (60 x 6) degrees in a circle, as well as the use of seconds and minutes of arc to denote fractions of a degree. It is likely the sexagesimal system was chosen because 60 can be evenly divided by 2, 3, 4, 5, 6, 10, 12, 15, 20 and 30. Also, unlike the Egyptians, Greeks, and Romans, the Babylonians had a true place-value system, where digits written in the left column represented larger values, much as in the decimal system. The power of the Babylonian notational system lay in that it could be used to represent fractions as easily as whole numbers; thus multiplying two numbers that contained fractions was no different than multiplying integers, similar to our modern notation. The notational system of the Babylonians was the best of any civilization until the Renaissance, and its power allowed it to achieve remarkable computation accuracy and power; for example, the Babylonian tablet YBC 7289 gives an approximation of "2 accurate to five decimal places. The Babylonians lacked, however, an equivalent of the decimal point, and so the place value of a symbol often had to be inferred from the context. By the Seleucid period, the Babylonians had developed a zero symbol as a placeholder for empty positions; however it was only used for intermediate positions. This zero sign does not appear in terminal positions, thus the Babylonians came close but did not develop a true place value system.

Other topics covered by Babylonian mathematics include fractions, algebra, quadratic and cubic equations, and the calculation of regular reciprocal pairs. The tablets also include multiplication tables and methods for solving linear, quadratic equations and cubic equations, a remarkable achievement for the time. Tablets from the Old Babylonian period also contain the earliest known statement of the Pythagorean theorem. However, as with Egyptian mathematics, Babylonian mathematics shows no awareness of the difference between exact and approximate solutions, or the solvability of a problem, and most importantly, no explicit statement of the need for proofs or logical principles.

Egyptian

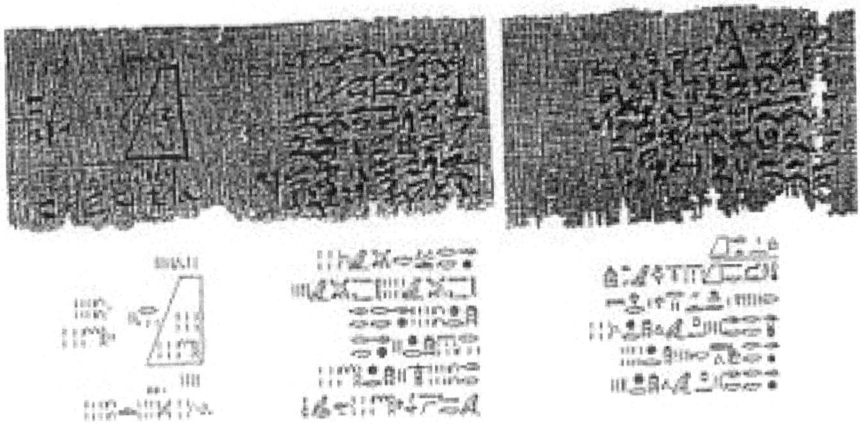

Image of Problem 14 from the Moscow Mathematical Papyrus. The problem includes a diagram indicating the dimensions of the truncated pyramid.

Egyptian mathematics refers to mathematics written in the Egyptian language. From the Hellenistic period, Greek replaced Egyptian as the written language of Egyptian scholars. Mathematical study in Egypt later continued under the Arab Empireas part of Islamic mathematics, when Arabic became the written language of Egyptian scholars.

The most extensive Egyptian mathematical text is the Rhind papyrus (sometimes also called the Ahmes Papyrus after its author), dated to c. 1650 BC but likely a copy of an older document from the Middle Kingdom of about 2000–1800 BC. It is an instruction manual for students in arithmetic and geometry. In addition to giving area formulas and methods for multiplication, division and working with unit fractions, it also contains evidence of other mathematical knowledge,including composite and prime numbers; arithmetic, geometric and harmonic means; and simplistic understandings of both the Sieve of Eratosthenes and perfect number theory (namely, that of the number 6). It also shows how to solve first order linear equations as well as arithmetic and geometric series.

Another significant Egyptian mathematical text is the Moscow papyrus, also from the Middle Kingdom period, dated to c. 1890 BC. It consists of what are today called word problems or story problems, which were apparently intended as entertainment. One problem is considered to be of particular importance because it gives a method for finding the volume of a frustum (truncated pyramid).

Finally, the Berlin Papyrus 6619 (c. 1800 BC) shows that ancient Egyptians could solve a second-order algebraic equation.

Greek

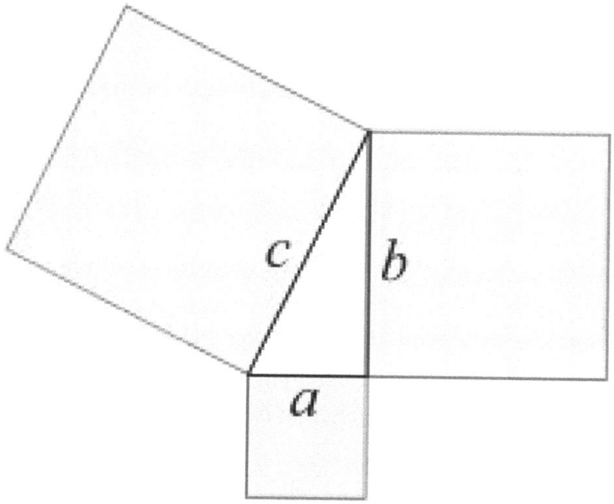

The Pythagorean theorem. The Pythagoreans are generally credited with the first proof of the theorem.

Greek mathematics refers to the mathematics written in the Greek language from the time of Thales of Miletus (~600 BC) to the closure of the Academy of Athens in 529 AD. Greek mathematicians lived in cities spread over the entire Eastern Mediterranean, from Italy to North Africa, but were united by culture and language. Greek mathematics of the period following Alexander the Great is sometimes called Hellenistic mathematics.

Greek mathematics was much more sophisticated than the mathematics that had been developed by earlier cultures. All surviving records of pre-Greek mathematics show the use of inductive reasoning, that is, repeated observations used to establish rules of thumb. Greek mathematicians, by contrast, used deductive reasoning. The Greeks used logic to derive conclusions from definitions and axioms, and used mathematical rigor to prove them.

Greek mathematics is thought to have begun with Thales of Miletus (c. 624–c.546 BC) and Pythagoras of Samos (c. 582–c. 507 BC). Although the extent of the influence is disputed, they were probably inspired by Egyptian and Babylonian mathematics. According to legend, Pythagoras traveled to Egypt to learn mathematics, geometry, and astronomy from Egyptian priests.

Thales used geometry to solve problems such as calculating the height of pyramids and the distance of ships from the shore. He is credited with the first use of deductive reasoning applied to geometry, by deriving four corollaries to Thales' Theorem. As a result, he has been hailed as the first true mathematician and the first known individual to whom a mathematical discovery has been attributed. Pythagoras established the Pythagorean School, whose doctrine it was that mathematics ruled the universe and whose motto was "All is number". It was the Pythagoreans who coined the term "mathematics", and with whom the study of mathematics for its own sake begins. The Pythagoreans are credited with the first proof of the Pythagorean theorem, though the statement of the theorem has a long history, and with the proof of the existence of irrational numbers. Although he was preceded by the Babylonians and the Chinese, the Neopythagorean mathematician Nicomachus (60–120 AD) provided one of the earliest Greco-Roman multiplication tables, whereas the oldest extant Greek multiplication table is found on a wax tablet dated to the 1st century AD (now found in the British Museum). The association of the Neopythagoreans with the Western invention of the multiplication table is evident in its later Medieval name: the mensa Pythagorica.

Plato (428/427 BC – 348/347 BC) is important in the history of mathematics for inspiring and guiding others. His Platonic Academy, in Athens, became the mathematical center of the world in the 4th century BC, and it was from this school that the leading mathematicians of the day, such as Eudoxus of Cnidus, came. Plato also discussed the foundations of mathematics, clarified some of the definitions (e.g. that of a line as "breadthless length"), and reorganized the assumptions. The analytic method is ascribed to Plato, while a formula for obtaining Pythagorean triples bears his name. Eudoxus (408–c. 355 BC) developed the method of exhaustion, a precursor of modern integration and a theory of ratios that avoided the problem of incommensurable magnitudes. The former allowed the calculations of areas and volumes of curvilinear figures, while the latter enabled subsequent geometers to make significant advances in geometry. Though he made no specific technical mathematical discoveries, Aristotle (384–c. 322 BC) contributed significantly to the development of mathematics by laying the foundations of logic.

One of the oldest surviving fragments of Euclid's Elements, found at Oxyrhynchus and dated to circa AD 100. The diagram accompanies Book II, Proposition 5.

In the 3rd century BC, the premier center of mathematical education and research was the Musaeum of Alexandria. It was there that Euclid (c. 300 BC) taught, and wrote the Elements, widely considered the most successful and influential textbook of all time. The Elementsintroduced mathematical rigor through the axiomatic method and is the earliest example of the format still used in mathematics today, that of definition, axiom, theorem, and proof. Although most of the contents of the Elements were already known, Euclid arranged them into a single, coherent logical framework. The Elements was known to all educated people in the West up through the middle of the 20th century and its contents are still taught in geometry classes today. In addition to the familiar theorems of Euclidean geometry, the Elements was meant as an introductory textbook to all mathematical subjects of the time, such as number theory, algebra and solid geometry, including proofs that the square root of two is irrational and that there are infinitely many prime numbers. Euclid also wrote extensively on other subjects, such as conic sections, optics, spherical geometry, and mechanics, but only half of his writings survive.

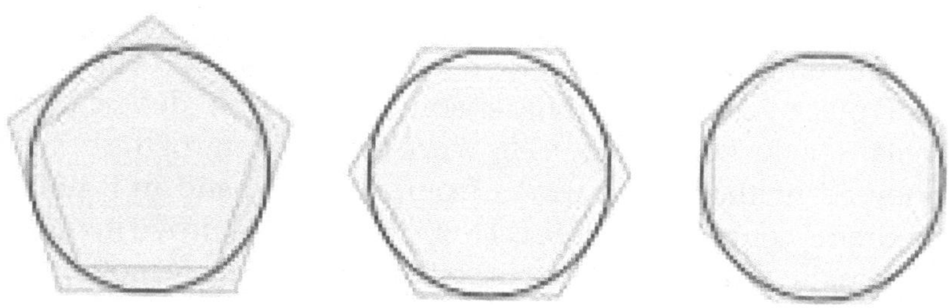

Archimedes used the method of exhaustionto approximate the value of pi.

Archimedes (c. 287–212 BC) of Syracuse, widely considered the greatest mathematician of antiquity, used the method of exhaustion to calculate the area under the arc of a parabolawith the summation of an infinite series, in a manner not too dissimilar from modern calculus. He also showed one could use the method of exhaustion to calculate the value of ð with as much precision as desired, and obtained the most accurate value of ð then known, 310/71 < ð < 310/70. He also studied the spiral bearing his name, obtained formulas for the volumes of surfaces of revolution (paraboloid, ellipsoid, hyperboloid), and an ingenious method of exponentiationfor expressing very large numbers. While he is also known for his contributions to physics and several advanced mechanical devices, Archimedes himself placed far greater value on the products of his thought and general mathematical principles. He regarded as his greatest achievement his finding of the surface area and volume of a sphere, which he obtained by proving these are 2/3 the surface area and volume of a cylinder circumscribing the sphere.

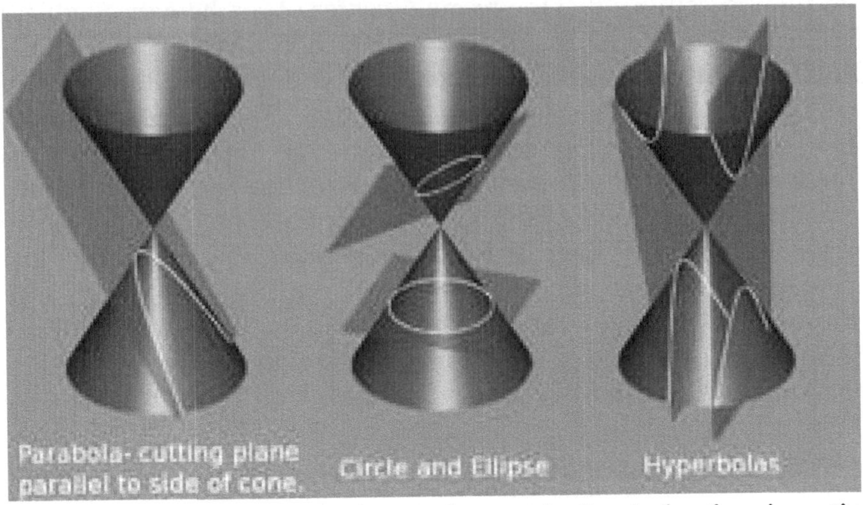

Apollonius of Perga made significant advances in the study of conic sections.

Apollonius of Perga (c. 262–190 BC) made significant advances to the study of conic sections, showing that one can obtain all three varieties of conic section by varying the angle of the plane that cuts a double-napped cone. He also coined the terminology in use today for conic sections, namely parabola ("place beside" or "comparison"), "ellipse" ("deficiency"), and "hyperbola" ("a throw beyond"). His work Conics is one of the best known and preserved mathematical works from antiquity, and in it he derives many theorems concerning conic sections that would prove invaluable to later mathematicians and astronomers studying planetary motion, such as Isaac Newton. While neither Apollonius nor any other Greek mathematicians made the leap to coordinate geometry, Apollonius' treatment of curves is in some ways similar to the modern treatment, and some of his work seems to anticipate the development of analytical geometry by Descartes some 1800 years later.

Around the same time, Eratosthenes of Cyrene (c. 276–194 BC) devised the Sieve of Eratosthenes for finding prime numbers.The 3rd century BC is generally regarded as the "Golden Age" of Greek mathematics, with advances in pure mathematics henceforth in relative decline. Nevertheless, in the centuries that followed significant advances were made in applied mathematics, most notably trigonometry, largely to address the needs of astronomers. Hipparchus of Nicaea (c. 190–120 BC) is considered the founder of trigonometry for compiling the first known trigonometric table, and to him is also due the systematic use of the 360 degree circle. Heron of Alexandria (c. 10–70 AD) is credited with Heron's formula for finding the area of a scalene triangle and with being the first to recognize the possibility of negative numbers possessing square roots. Menelaus of Alexandria (c.

100 AD) pioneered spherical trigonometry through Menelaus' theorem. The most complete and influential trigonometric work of antiquity is the Almagest of Ptolemy (c. AD 90–168), a landmark astronomical treatise whose trigonometric tables would be used by astronomers for the next thousand years. Ptolemy is also credited with Ptolemy's theorem for deriving trigonometric quantities, and the most accurate value of ð outside of China until the medieval period, 3.1416.

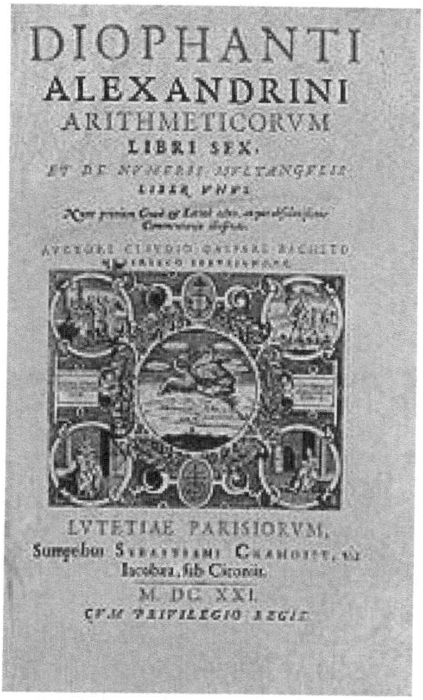

Title page of the 1621 edition of Diophantus' Arithmetica, translated into Latin by Claude Gaspard Bachet de Méziriac.

Following a period of stagnation after Ptolemy, the period between 250 and 350 AD is sometimes referred to as the "Silver Age" of Greek mathematics. During this period, Diophantus made significant advances in algebra, particularly indeterminate analysis, which is also known as "Diophantine analysis". The study of Diophantine equations and Diophantine approximations is a significant area of research to this day. His main work was the Arithmetica, a collection of 150 algebraic problems dealing with exact solutions to determinate and indeterminate equations. The Arithmetica had a significant influence on later mathematicians, such as Pierre de Fermat, who arrived at his famous Last Theorem after trying to generalize a problem he had read in the Arithmetica (that of dividing a square into two squares). Diophantus also made significant advances in notation, the Arithmetica being the first instance of algebraic symbolism and syncopation.

The Hagia Sophia was designed by mathematicians Anthemius of Trallesand Isidore of Miletus.

Among the last great Greek mathematicians is Pappus of Alexandria (4th century AD). He is known for his hexagon theorem and centroid theorem, as well as the Pappus configuration and Pappus graph. His Collection is a major source of knowledge on Greek mathematics as most of it has survived. Pappus is considered the last major innovator in Greek mathematics, with subsequent work consisting mostly of commentaries on earlier work.

The first woman mathematician recorded by history was Hypatia of Alexandria (AD 350–415). She succeeded her father (Theon of Alexandria) as Librarian at the Great Library and wrote many works on applied mathematics. Because of a political dispute, the Christian community in Alexandria had her stripped publicly and executed. Her death is sometimes taken as the end of the era of the Alexandrian Greek mathematics, although work did continue in Athens for another century with figures such as Proclus, Simplicius and Eutocius. Although Proclus and Simplicius were more philosophers than mathematicians, their commentaries on earlier works are valuable sources on Greek mathematics. The closure of the neo-Platonic Academy of Athens by the emperor Justinian in 529 AD is traditionally held as marking the end of the era of Greek mathematics, although the Greek tradition continued unbroken in the Byzantine empire with mathematicians such as Anthemius of Tralles and Isidore of Miletus, the architects of the Hagia Sophia. Nevertheless, Byzantine mathematics consisted mostly of commentaries, with little in the way of innovation, and the centers of mathematical innovation were to be found elsewhere by this time.

Roman

Equipment used by an ancient Roman land surveyor (gromatici), found at the site of Aquincum, modern Budapest, Hungary

Although ethnic Greek mathematicians continued under the rule of the late Roman Republic and subsequent Roman Empire, there were no noteworthy native Latin mathematicians in comparison. Ancient Romans such as Cicero (106–43 BC), an influential Roman statesman who studied mathematics in Greece, believed that Roman surveyors and calculators were far more interested in applied mathematics than the theoretical mathematics and geometry that were prized by the Greeks. It is unclear if the Romans first derived their numerical systemdirectly from the Greek precedent or from Etruscan numerals used by the Etruscan civilization centered in what is now Tuscany, central Italy.

Using calculation, Romans were adept at both instigating and detecting financial fraud, as well as managing taxes for the treasury. Siculus Flaccus, one of the Roman gromatici (i.e. land surveyor), wrote the Categories of Fields, which aided Roman surveyors in measuring the surface areas of allotted lands and territories. Aside from managing trade and taxes, the Romans also regularly applied mathematics to solve problems in engineering, including the erection of architecture such as bridges, road-building, and preparation for military campaigns. Arts and crafts such as Roman mosaics, inspired by previous Greek designs, created illusionist geometric patterns and rich, detailed scenes that required precise measurements for each tessera tile, the opus tessellatum pieces on average measuring eight millimeters square and the finer opus vermiculatum pieces having an average surface of four millimeters square.

The creation of the Roman calendar also necessitated basic mathematics. The first calendar allegedly dates back to 8th century BC during the Roman Kingdom and included 356 days plus a leap year every other year. In contrast, the lunar calendar of the Republican era contained 355 days, roughly ten-and-one-fourth days shorter than the solar year, a discrepancy that was solved by adding an extra month into the calendar after the 23rd of February. This calendar was supplanted by the Julian calendar, a solar calendar organized by Julius Caesar (100–44 BC) and devised by Sosigenes of Alexandria to include a leap day every four years in a 365-day cycle. This calendar, which contained an error of 11 minutes and 14 seconds, was later corrected by the Gregorian calendar organized by Pope Gregory XIII (r. 1572–1585), virtually the same solar calendar used in modern times as the international standard calendar.

At roughly the same time, the Han Chinese and the Romans both invented the wheeled odometer device for measuring distances traveled, the Roman model first described by the Roman civil engineer and architect Vitruvius (c. 80 BC - c. 15 BC). The device was used at least until the reign of emperor Commodus (r. 177 – 192 AD), but its design seems to have been lost until experiments were made during the 15th century in Western Europe. Perhaps relying on similar gear-work and technology found in the Antikythera mechanism, the odometer of Vitruvius featured chariot wheels measuring 4 feet (1.2 m) in diameter turning four-hundred times in one Roman mile (roughly 4590 ft/1400 m). With each revolution, a pin-and-axle device engaged a 400-tooth cogwheel that turned a second gear responsible for dropping pebbles into a box, each pebble representing one mile traversed.

Chinese[edit]

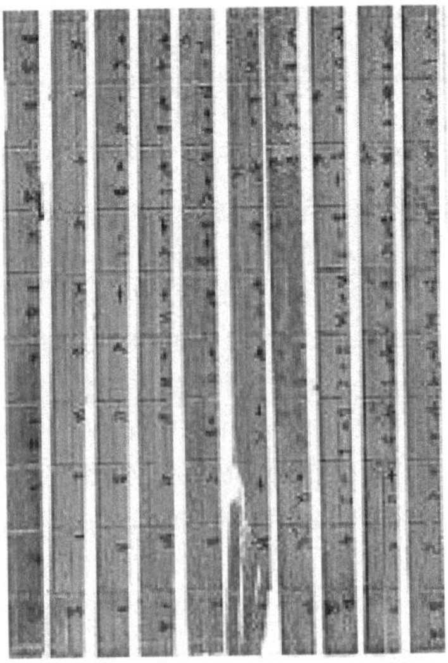

The Tsinghua Bamboo Slips, containing the world's earliest decimal multiplication table, dated 305 BC during the Warring States period

An analysis of early Chinese mathematics has demonstrated its unique development compared to other parts of the world, leading scholars to assume an entirely independent development. The oldest extant mathematical text from China is the Zhoubi Suanjing, variously dated to between 1200 BC and 100 BC, though a date of about 300 BC during the Warring States Period appears reasonable. However, the Tsinghua Bamboo Slips, containing the earliest known decimal multiplication table (although ancient Babylonians had ones with a base of 60), is dated around 305 BC and is perhaps the oldest surviving mathematical text of China.

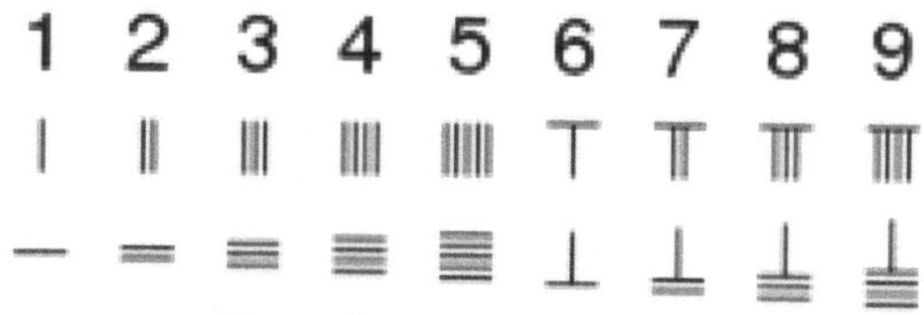

Counting rod numerals

Of particular note is the use in Chinese mathematics of a decimal positional notation system, the so-called "rod numerals" in which distinct ciphers were used for numbers between 1 and 10, and additional ciphers for powers of ten. Thus, the number 123 would be written using the symbol for "1", followed by the symbol for "100", then the symbol for "2" followed by the symbol for "10", followed by the symbol for "3". This was the most advanced number system in the world at the time, apparently in use several centuries before the common era and well before the development of the Indian numeral system. Rod numerals allowed the representation of numbers as large as desired and allowed calculations to be carried out on the suan pan, or Chinese abacus. The date of the invention of the suan pan is not certain, but the earliest written mention dates from AD 190, in Xu Yue's Supplementary Notes on the Art of Figures.

The oldest existent work on geometry in China comes from the philosophical Mohist canon c. 330 BC, compiled by the followers of Mozi (470–390 BC). The Mo Jing described various aspects of many fields associated with physical science, and provided a small number of geometrical theorems as well. It also defined the concepts of circumference, diameter, radius, and volume.

In 212 BC, the Emperor Qin Shi Huang commanded all books in the Qin Empire other than officially sanctioned ones be burned. This decree was not universally obeyed, but as a consequence of this order little is known about ancient Chinese mathematics before this date. After the book burning of 212 BC, the Han dynasty (202 BC–220 AD) produced works of mathematics which presumably expanded on works that are now lost. The most important of these is The Nine Chapters on the Mathematical Art, the full title of which appeared by AD 179, but existed in part under other titles beforehand. It consists of 246 word problems involving agriculture, business, employment of geometry to figure height spans and dimension ratios for Chinese pagoda towers, engineering, surveying, and includes material on right triangles. It created mathematical proof for the Pythagorean theorem, and a mathematical formula for Gaussian elimination. The treatise also provides values of ð, which Chinese mathematicians originally approximated as 3 until Liu Xin (d. 23 AD) provided a figure of 3.1457 and subsequently Zhang Heng (78–139) approximated pi as 3.1724, as well as 3.162 by taking the square root of 10. Liu Hui commented on the Nine Chapters in the 3rd century AD and gave a value of ð accurate to 5 decimal places (i.e. 3.14159). Though more of a matter of computational stamina than theoretical insight, in the 5th century AD Zu Chongzhi computed the value of ð to seven decimal places (i.e. 3.141592), which remained the most accurate value of ð for almost the next 1000 years. He also established a method which would later be called Cavalieri's principle to find the volume of a sphere.

The high-water mark of Chinese mathematics occurred in the 13th century during the latter half of the Song dynasty (960–1279), with the development of Chinese algebra. The most important text from that period is the Precious Mirror of the Four Elements by Zhu Shijie (1249–1314), dealing with the solution of simultaneous higher order algebraic equations using a method similar to Horner's method. The Precious Mirror also contains a diagram of Pascal's triangle with coefficients of binomial expansions through the eighth power, though both appear in Chinese works as early as 1100. The Chinese also made use of the complex combinatorial diagram known as the magic square and magic circles, described in ancient times and perfected by Yang Hui (AD 1238–1298).

Even after European mathematics began to flourish during the Renaissance, European and Chinese mathematics were separate traditions, with significant Chinese mathematical output in decline from the 13th century onwards. Jesuit missionaries such as Matteo Ricci carried mathematical ideas back and forth between the two cultures from the 16th to 18th centuries, though at this point far more mathematical ideas were entering China than leaving.

Japanese mathematics, Korean mathematics, and Vietnamese mathematics are traditionally viewed as stemming from Chinese mathematics and belonging to the Confucian-based East Asian cultural sphere. Korean and Japanese mathematics were heavily influenced by the algebraic works produced during China's Song dynasty, whereas Vietnamese mathematics was heavily indebted to popular works of China's

Ming dynasty (1368–1644). For instance, although Vietnamese mathematical treatises were written in either Chinese or the native Vietnamese Chï Nôm script, all of them followed the Chinese format of presenting a collection of problems with algorithms for solving them, followed by numerical answers. Mathematics in Vietnam and Korea were mostly associated with the professional court bureaucracy of mathematicians and astronomers, whereas in Japan it was more prevalent in the realm of private schools.

Indian

The numerals used in the Bakhshali manuscript, dated between the 2nd century BCE and the 2nd century CE.

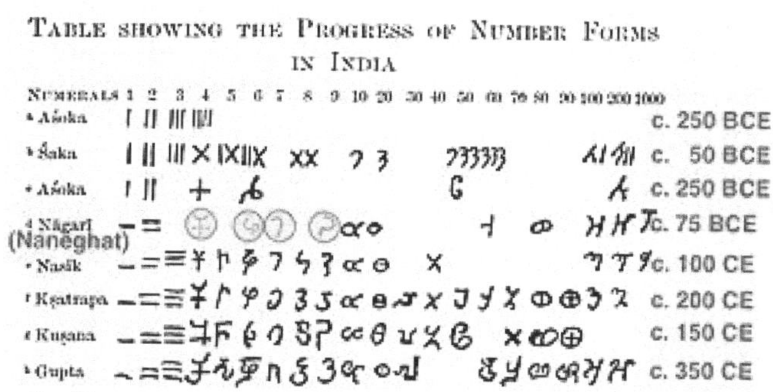

Indian numerals in stone and copper inscriptions

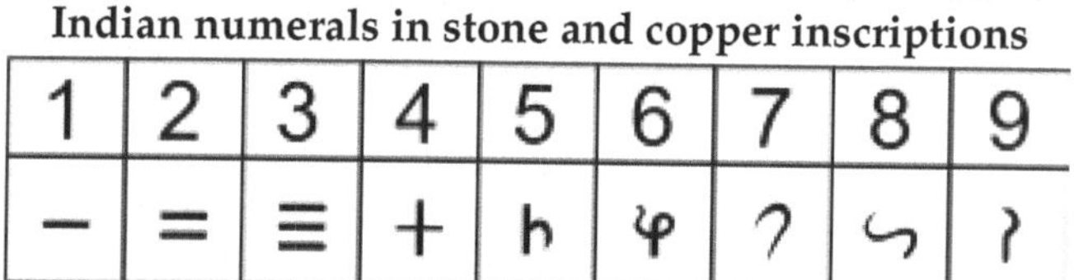

Ancient Brahmi numerals in a part of India

The earliest civilization on the Indian subcontinent is the Indus Valley Civilization (mature phase: 2600 to 1900 BC) that flourished in the Indus river basin. Their cities were laid out with geometric regularity, but no known mathematical documents survive from this civilization.

The oldest extant mathematical records from India are the Sulba Sutras (dated variously between the 8th century BC and the 2nd century AD), appendices to religious texts which give simple rules for constructing altars of various shapes, such as squares, rectangles, parallelograms, and others. As with Egypt, the preoccupation with temple functions points to an origin of mathematics in religious ritual. The Sulba Sutras give methods for constructing a circle with approximately the same area as a given square, which imply several different approximations of the value of ð.In addition, they compute the square root of 2 to several decimal places, list Pythagorean triples, and give a statement of the Pythagorean theorem. All of these results are present in Babylonian mathematics, indicating Mesopotamian influence.

It is not known to what extent the Sulba Sutras influenced later Indian mathematicians. As in China, there is a lack of continuity in Indian mathematics; significant advances are separated by long periods of inactivity.

PâGini (c. 5th century BC) formulated the rules for Sanskrit grammar. His notation was similar to modern mathematical notation, and used metarules, transformations, and recursion. Pingala (roughly 3rd–1st centuries BC) in his treatise of prosody uses a device corresponding to a binary numeral system. His discussion of the combinatorics of meterscorresponds to an elementary version of the binomial theorem. Pingala's work also contains the basic ideas of Fibonacci numbers (called mâtrâmeru).

The next significant mathematical documents from India after the Sulba Sutras are the Siddhantas, astronomical treatises from the 4th and 5th centuries AD (Gupta period) showing strong Hellenistic influence. They are significant in that they contain the first instance of trigonometric relations based on the half-chord, as is the case in modern trigonometry, rather than the full chord, as was the case in Ptolemaic trigonometry. Through a series of translation errors, the words "sine" and "cosine" derive from the Sanskrit "jiya" and "kojiya".

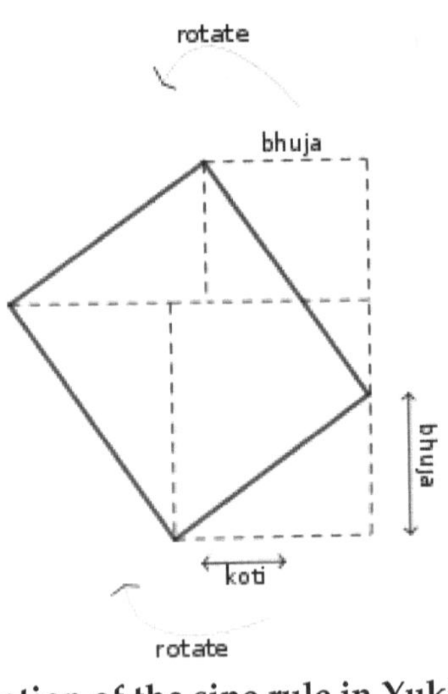

Explanation of the sine rule in Yuktibhâcâ

Around 500 AD, Aryabhata wrote the Aryabhatiya, a slim volume, written in verse, intended to supplement the rules of calculation used in astronomy and mathematical mensuration, though with no feeling for logic or deductive methodology. Though about half of the entries are wrong, it is in the Aryabhatiya that the decimal place-value system first appears. Several centuries later, the Muslim mathematician Abu Rayhan

Biruni described the Aryabhatiya as a "mix of common pebbles and costly crystals".

In the 7th century, Brahmagupta identified the Brahmagupta theorem, Brahmagupta's identity and Brahmagupta's formula, and for the first time, in Brahma-sphuta-siddhanta, he lucidly explained the use of zero as both a placeholder and decimal digit, and explained the Hindu–Arabic numeral system. It was from a translation of this Indian text on mathematics (c. 770) that Islamic mathematicians were introduced to this numeral system, which they adapted as Arabic numerals. Islamic scholars carried knowledge of this number system to Europe by the 12th century, and it has now displaced all older number systems throughout the world. Various symbol sets are used to represent numbers in the Hindu–Arabic numeral system, all of which evolved from the Brahmi numerals. Each of the roughly dozen major scripts of India has its own numeral glyphs. In the 10th century, Halayudha's commentary on Pingala's work contains a study of the Fibonacci sequence and Pascal's triangle, and describes the formation of a matrix.

In the 12th century, Bhâskara II lived in southern India and wrote extensively on all then known branches of mathematics. His work contains mathematical objects equivalent or approximately equivalent to infinitesimals, derivatives, the mean value theorem and the derivative of the sine function. To what extent he anticipated the invention of calculus is a controversial subject among historians of mathematics.

In the 14th century, Madhava of Sangamagrama, the founder of the so-called Kerala School of Mathematics, found the Madhava–Leibniz series, and, using 21 terms, computed the value of ð as 3.14159265359. Madhava also found the Madhava-Gregory series to determine the arctangent, the Madhava-Newton power series to determine sine and cosine and the Taylor approximation for sine and cosine functions. In the 16th century, Jyesthadeva consolidated many of the Kerala School's developments and theorems in the Yukti-bhâcâ. However, the Kerala School did not formulate a systematic theory of differentiation and integration, nor is there any direct evidence of their results being transmitted outside Kerala.

Islamic empire

The Islamic Empire established across Persia, the Middle East, Central Asia, North Africa, Iberia, and in parts of India in the 8th century made significant contributions towards mathematics. Although most Islamic texts on mathematics were written in Arabic, most of them were not written by Arabs, since much like the status of Greek in the Hellenistic world, Arabic was used as the written language of non-Arab scholars throughout the Islamic world at the time. Persians contributed to the world of Mathematics alongside Arabs.

In the 9th century, the Persian mathematician Mu%ammad ibn Mûsâ al-Khwârizmî wrote several important books on the Hindu–Arabic numerals and on methods for solving equations. His book On the Calculation with Hindu Numerals, written about 825, along with the work of Al-Kindi, were instrumental in spreading Indian mathematics and Indian numerals to the West. The word algorithm is derived from the Latinization of his name, Algoritmi, and the word algebra from the title of one of his works, Al-Kitâb al-mukhtacar fî hîsâb al-ðabr wa'l-muqâbala (The Compendious Book on Calculation by Completion and Balancing). He gave an exhaustive explanation for the algebraic solution of quadratic equations with positive roots, and he was the first to teach algebra in an elementary form and for its own sake.He also discussed the fundamental method of "reduction" and "balancing", referring to the transposition of subtracted terms to the other side of an equation, that is, the cancellation of like terms on opposite sides of the equation. This is the operation which al-Khwârizmî originally described as al-jabr. His algebra was also no longer concerned "with a series of problems to be resolved, but an exposition which starts with primitive terms in which the combinations must give all possible prototypes for equations, which henceforward explicitly constitute the true object of study." He also studied an equation for its own sake and "in a generic manner, insofar as it does not simply emerge in the course of solving a problem, but is specifically called on to define an infinite class of problems."

In Egypt, Abu Kamil extended algebra to the set of irrational numbers, accepting square roots and fourth roots as solutions and coefficients to quadratic equations. He also developed techniques used to solve three non-linear simultaneous equations with three unknown variables. One unique feature of his works was trying to find all the possible solutions to some of his problems, including one where he found 2676 solutions. His works formed an important foundation for the development of algebra and influenced later mathematicians, such as al-Karaji and Fibonacci.

Further developments in algebra were made by Al-Karaji in his treatise al-Fakhri, where he extends the methodology to incorporate integer powers and integer roots of unknown quantities. Something close to a proof by mathematical induction appears in a book written by Al-Karaji around 1000 AD, who used it to prove the binomial theorem, Pascal's triangle, and the sum of integral cubes. The historian of mathematics, F. Woepcke, praised Al-Karaji for being "the first who introduced the theory of algebraic calculus." Also in the 10th century, Abul Wafa translated the works of Diophantus into Arabic. Ibn al-Haytham was the first mathematician to derive the formula for the sum of the fourth powers, using a method that is readily generalizable for determining the general formula for the sum of any integral powers. He performed an integration in order to find the volume of a paraboloid, and was able to generalize his result for the integrals of polynomials up to the fourth degree. He thus came close to finding a general formula for the integrals of polynomials, but he was not concerned with any polynomials higher than the fourth degree.

In the late 11th century, Omar Khayyam wrote Discussions of the Difficulties in Euclid, a book about what he perceived as flaws in Euclid's Elements, especially the parallel postulate. He was also the first to find the general geometric solution to cubic equations. He was also very influential in calendar reform.

In the 13th century, Nasir al-Din Tusi (Nasireddin) made advances in spherical trigonometry. He also wrote influential work on Euclid's parallel postulate. In the 15th century, Ghiyath al-Kashi computed the value of ð to the 16th decimal place. Kashi also had an algorithm for calculating nth roots, which was a special case of the methods given many centuries later by Ruffini and Horner.

Other achievements of Muslim mathematicians during this period include the addition of the decimal point notation to the Arabic numerals, the discovery of all the modern trigonometric functions besides the sine, al-Kindi's introduction of cryptanalysis and frequency analysis, the development of analytic geometry by Ibn al-Haytham, the beginning of algebraic geometry by Omar Khayyam and the development of an algebraic notation by al-Qalasâdî.

During the time of the Ottoman Empire and Safavid Empire from the 15th century, the development of Islamic mathematics became stagnant.

Maya

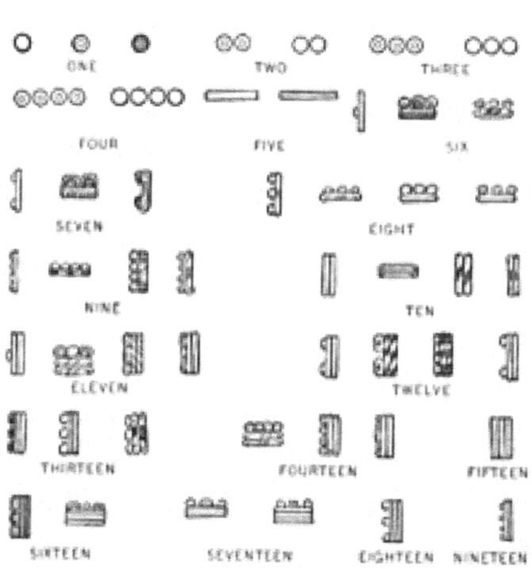

The Maya numerals for numbers 1 through 19, written in the Maya script

In the Pre-Columbian Americas, the Maya civilization that flourished in Mexico and Central America during the 1st millennium AD developed a unique tradition of mathematics that, due to its geographic isolation, was entirely independent of existing European, Egyptian, and Asian mathematics. Maya numerals utilized a base of 20, the vigesimal system, instead of a base of ten that forms the basis of the decimalsystem used by most modern cultures. The Mayas used mathematics to create the Maya calendar as well as to predict astronomical phenomena in their native Maya astronomy. While the concept of zero had to be inferred in the mathematics of many contemporary cultures, the Mayas developed a standard symbol for it.

MEDIEVAL EUROPEAN

Medieval European interest in mathematics was driven by concerns quite different from those of modern mathematicians. One driving element was the belief that mathematics provided the key to understanding the created order of nature, frequently justified by Plato's Timaeus and the biblical passage (in the Book of Wisdom) that God had ordered all things in measure, and number, and weight.

Boethius provided a place for mathematics in the curriculum in the 6th century when he coined the term quadrivium to describe the study of arithmetic, geometry, astronomy, and music. He wrote De institutione arithmetica, a free translation from the Greek of Nicomachus's Introduction to Arithmetic; De institutione musica, also derived from Greek sources; and a series of excerpts from Euclid's Elements. His works were theoretical, rather than practical, and were the basis of mathematical study until the recovery of Greek and Arabic mathematical works.

In the 12th century, European scholars traveled to Spain and Sicily seeking scientific Arabic texts, including al-Khwârizmî's The Compendious Book on Calculation by Completion and Balancing, translated into Latin by Robert of Chester, and the complete text of Euclid's Elements, translated in various versions by Adelard of Bath, Herman of Carinthia, and Gerard of Cremona. These and other new sources sparked a renewal of mathematics.

Leonardo of Pisa, now known as Fibonacci, serendipitously learned about the Hindu–Arabic numerals on a trip to what is now Béjaïa, Algeria with his merchant father. (Europe was still using Roman numerals.) There, he observed a system of arithmetic (specifically algorism) which due to the positional notation of Hindu–Arabic numerals was much more efficient and greatly facilitated commerce. Leonardo wrote Liber Abaci in 1202 (updated in 1254) introducing the technique to Europe and beginning a long period of popularizing it. The book also brought to Europe what is now known as the Fibonacci sequence (known to Indian mathematicians for hundreds of years before that) which was used as an unremarkable example within the text.

The 14th century saw the development of new mathematical concepts to investigate a wide range of problems. One important contribution was development of mathematics of local motion.

Thomas Bradwardine proposed that speed (V) increases in arithmetic proportion as the ratio of force (F) to resistance (R) increases in geometric proportion. Bradwardine expressed this by a series of specific examples, but although the logarithm had not yet been conceived, we can express his conclusion anachronistically by writing: $V = \log (F/R)$.Bradwardine's analysis is an example of transferring a mathematical technique used by al-Kindi and Arnald of Villanova to quantify the nature of compound medicines to a different physical problem.

Nicole Oresme (1323–1382), shown in this contemporary illuminated manuscript with an armillary sphere in the foreground, was the first to offer a mathematical proof for the divergenceof the harmonic series.

One of the 14th-century Oxford Calculators, William Heytesbury, lacking differential calculus and the concept of limits, proposed to measure instantaneous speed "by the path that would be described by [a body] if... it were moved uniformly at the same degree of speed with which it is moved in that given instant".

Heytesbury and others mathematically determined the distance covered by a body undergoing uniformly accelerated motion (today solved by integration), stating that "a moving body uniformly acquiring or losing that increment [of speed] will traverse in some given time a [distance] completely equal to that which it would traverse if it were moving continuously through the same time with the mean degree [of speed]".

Nicole Oresme at the University of Paris and the Italian Giovanni di Casali independently provided graphical demonstrations of this relationship, asserting that the area under the line depicting the constant acceleration, represented the total distance traveled. In a later mathematical commentary on Euclid's Elements, Oresme made a more detailed general analysis in which he demonstrated that a body will acquire in each successive increment of time an increment of any quality that increases as the odd numbers. Since Euclid had demonstrated the sum of the odd numbers are the square numbers, the total quality acquired by the body increases as the square of the time.

Renaissance

During the Renaissance, the development of mathematics and of accounting were intertwined. While there is no direct relationship between algebra and accounting, the teaching of the subjects and the books published often intended for the children of merchants who were sent to reckoning schools (in Flandersand Germany) or abacus schools (known as abbaco in Italy), where they learned the skills useful for trade and commerce. There is probably no need for algebra in performing bookkeeping operations, but for complex bartering operations or the calculation of compound interest, a basic knowledge of arithmetic was mandatory and knowledge of algebra was very useful.

Piero della Francesca (c. 1415–1492) wrote books on solid geometry and linear perspective, including De Prospectiva Pingendi (On Perspective for Painting), Trattato d'Abaco (Abacus Treatise), and De corporibus regularibus (Regular Solids).

Luca Pacioli's Summa de Arithmetica, Geometria, Proportioni et Proportionalità (Italian: "Review of Arithmetic, Geometry, Ratio and Proportion") was first printed and published in Venice in 1494. It included a 27-page treatise on bookkeeping, "Particularis de Computis et Scripturis" (Italian: "Details of Calculation and Recording"). It was written primarily for, and sold mainly to, merchants who used the book as a reference text, as a source of pleasure from the mathematical puzzles it contained, and to aid the education of their sons. In Summa Arithmetica, Pacioli introduced symbols for plus and minus for the first time in a printed book, symbols that became standard notation in Italian Renaissance mathematics. Summa Arithmetica was also the first known book printed in Italy to contain algebra. Pacioli obtained many of his ideas from Piero Della Francesca whom he plagiarized.

In Italy, during the first half of the 16th century, Scipione del Ferro and Niccolò Fontana Tartaglia discovered solutions for cubic equations. Gerolamo Cardano published them in his 1545 book Ars Magna, together with a solution for the quartic equations, discovered by his student Lodovico Ferrari. In 1572 Rafael Bombelli published his L'Algebra in which he showed how to deal with the imaginary quantities that could appear in Cardano's formula for solving cubic equations.

Simon Stevin's book De Thiende ('the art of tenths'), first published in Dutch in 1585, contained the first systematic treatment of decimal notation, which influenced all later work on the real number system.

Driven by the demands of navigation and the growing need for accurate maps of large areas, trigonometry grew to be a major branch of mathematics. Bartholomaeus Pitiscus was the first to use the word, publishing his Trigonometria in 1595. Regiomontanus's table of sines and cosines was published in 1533.

During the Renaissance the desire of artists to represent the natural world realistically, together with the rediscovered philosophy of the Greeks, led artists to study mathematics. They were also the engineers and architects of that time, and so had need of mathematics in any case. The art of painting in perspective, and the developments in geometry that involved, were studied intensely.

Mathematics during the Scientific Revolution

17th century

Gottfried Wilhelm Leibniz.

The 17th century saw an unprecedented increase of mathematical and scientific ideas across Europe. Galileo observed the moons of Jupiter in orbit about that planet, using a telescope based on a toy imported from Holland. Tycho Brahe had gathered an enormous quantity of mathematical data describing the positions of the planets in the sky. By his position as Brahe's assistant, Johannes Kepler was first exposed to and seriously interacted with the topic of planetary motion. Kepler's calculations were made simpler by the contemporaneous invention of logarithms by John Napier and Jost Bürgi. Kepler succeeded in formulating mathematical laws of planetary motion. The analytic geometry developed by René Descartes (1596–1650) allowed those orbits to be plotted on a graph, in Cartesian coordinates.

Building on earlier work by many predecessors, Isaac Newton discovered the laws of physics explaining Kepler's Laws, and brought together the concepts now known as calculus. Independently, Gottfried Wilhelm Leibniz, who is arguably one of the most important mathematicians of the 17th century, developed calculus and much of the calculus notation still in use today. Science and mathematics had become an international endeavor, which would soon spread over the entire world.

In addition to the application of mathematics to the studies of the heavens, applied mathematics began to expand into new areas, with the correspondence of Pierre de Fermat and Blaise Pascal. Pascal and Fermat set the groundwork for the investigations of probability theory and the corresponding rules of combinatorics in their discussions over a game of gambling. Pascal, with his wager, attempted to use the newly developing probability theory to argue for a life devoted to religion, on the grounds that even if the probability of success was small, the rewards were infinite. In some sense, this foreshadowed the development of utility theory in the 18th–19th century.

18th century

The most influential mathematician of the 18th century was arguably Leonhard Euler. His contributions range from founding the study of graph theorywith the Seven Bridges of Königsberg problem to standardizing many modern mathematical terms and notations. For example, he named the square root of minus 1 with the symbol i, and he popularized the use of the Greek letter to stand for the ratio of a circle's circumference to its diameter. He made numerous contributions to the study of topology, graph theory, calculus, combinatorics, and complex analysis, as evidenced by the multitude of theorems and notations named for him.

Other important European mathematicians of the 18th century included Joseph Louis Lagrange, who did pioneering work in number theory, algebra, differential calculus, and the calculus of variations, and Laplace who, in the age of Napoleon, did important work on the foundations of celestial mechanics and on statistics.

Modern

19th century

Carl Friedrich Gauss.

Throughout the 19th century mathematics became increasingly abstract. Carl Friedrich Gauss (1777–1855) epitomizes this trend. He did revolutionary work on functions of complex variables, in geometry, and on the convergence of series, leaving aside his many contributions to science. He also gave the first satisfactory proofs of the fundamental theorem of algebra and of the quadratic reciprocity law.

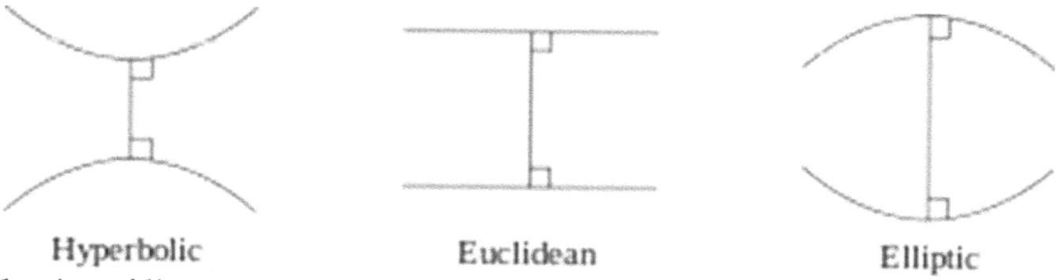

Behavior of lines with a common perpendicular in each of the three types of geometry

This century saw the development of the two forms of non-Euclidean geometry, where the parallel postulate of Euclidean geometry no longer holds. The Russian mathematician Nikolai Ivanovich Lobachevsky and his rival, the Hungarian mathematician János Bolyai, independently defined and studied hyperbolic geometry, where uniqueness of parallels no longer holds. In this geometry the sum of angles in a triangle add up to less than 180°. Elliptic geometry was developed later in the 19th century by the

German mathematician Bernhard Riemann; here no parallel can be found and the angles in a triangle add up to more than 180°. Riemann also developed Riemannian geometry, which unifies and vastly generalizes the three types of geometry, and he defined the concept of a manifold, which generalizes the ideas of curves and surfaces.

The 19th century saw the beginning of a great deal of abstract algebra. Hermann Grassmann in Germany gave a first version of vector spaces, William Rowan Hamilton in Ireland developed noncommutative algebra. The British mathematician George Boole devised an algebra that soon evolved into what is now called Boolean algebra, in which the only numbers were 0 and 1. Boolean algebra is the starting point of mathematical logic and has important applications in computer science.

Augustin-Louis Cauchy, Bernhard Riemann, and Karl Weierstrass reformulated the calculus in a more rigorous fashion.

Also, for the first time, the limits of mathematics were explored. Niels Henrik Abel, a Norwegian, and Évariste Galois, a Frenchman, proved that there is no general algebraic method for solving polynomial equations of degree greater than four (Abel–Ruffini theorem). Other 19th-century mathematicians utilized this in their proofs that straightedge and compass alone are not sufficient to trisect an arbitrary angle, to construct the side of a cube twice the volume of a given cube, nor to construct a square equal in area to a given circle. Mathematicians had vainly attempted to solve all of these problems since the time of the ancient Greeks. On the other hand, the limitation of three dimensions in geometry was surpassed in the 19th century through considerations of parameter space and hypercomplex numbers.

Abel and Galois's investigations into the solutions of various polynomial equations laid the groundwork for further developments of group theory, and the associated fields of abstract algebra. In the 20th century physicists and other scientists have seen group theory as the ideal way to study symmetry.

In the later 19th century, Georg Cantor established the first foundations of set theory, which enabled the rigorous treatment of the notion of infinity and has become the common language of nearly all mathematics. Cantor's set theory, and the rise of mathematical logic in the hands of Peano, L.E.J. Brouwer, David Hilbert, Bertrand Russell, and A.N. Whitehead, initiated a long running debate on the foundations of mathematics.

The 19th century saw the founding of a number of national mathematical societies: the London Mathematical Society in 1865, the Société Mathématique de France in 1872, the Circolo Matematico di Palermo in 1884, the Edinburgh Mathematical Society in 1883, and the American

Mathematical Society in 1888. The first international, special-interest society, the Quaternion Society, was formed in 1899, in the context of a vector controversy.

In 1897, Hensel introduced p-adic numbers.

20th century

The 20th century saw mathematics become a major profession. Every year, thousands of new Ph.D.s in mathematics were awarded, and jobs were available in both teaching and industry. An effort to catalogue the areas and applications of mathematics was undertaken in Klein's encyclopedia.

In a 1900 speech to the International Congress of Mathematicians, David Hilbert set out a list of 23 unsolved problems in mathematics. These problems, spanning many areas of mathematics, formed a central focus for much of 20th-century mathematics. Today, 10 have been solved, 7 are partially solved, and 2 are still open. The remaining 4 are too loosely formulated to be stated as solved or not.

A map illustrating the Four Color Theorem

Notable historical conjectures were finally proven. In 1976, Wolfgang Haken and Kenneth Appel proved the four color theorem, controversial at the time for the use of a computer to do so. Andrew Wiles, building on the work of others, proved Fermat's Last Theorem in 1995. Paul Cohen and Kurt Gödel proved that the continuum hypothesis is independent of (could neither be proved nor disproved from) the standard axioms of set theory. In 1998 Thomas Callister Hales proved the Kepler conjecture.

Mathematical collaborations of unprecedented size and scope took place. An example is the classification of finite simple groups (also called the "enormous theorem"), whose proof between 1955 and 1983 required 500-odd journal articles by about 100 authors, and filling tens of thousands of pages. A group of French mathematicians, including Jean Dieudonné and André Weil, publishing under the pseudonym "Nicolas Bourbaki", attempted to exposit all of known mathematics as a coherent rigorous whole. The resulting several dozen volumes has had a controversial influence on mathematical education.

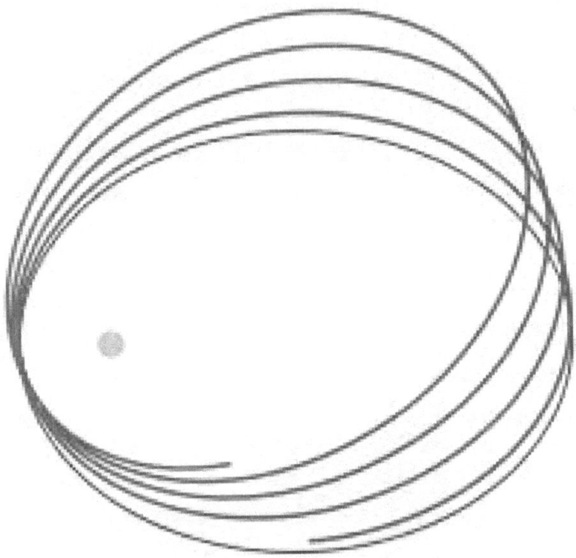

Newtonian (red) vs. Einsteinian orbit (blue) of a lone planet orbiting a star, with relativistic precession of apsides

Differential geometry came into its own when Einstein used it in general relativity. Entirely new areas of mathematics such as mathematical logic, topology, and John von Neumann's game theory changed the kinds of questions that could be answered by mathematical methods. All kinds of structures were abstracted using axioms and given names like metric spaces, topological spaces etc. As mathematicians do, the concept of an abstract structure was itself abstracted and led to category theory. Grothendieck and Serrerecast algebraic geometry using sheaf theory. Large advances were made in the qualitative study of dynamical systems that Poincaré had begun in the 1890s. Measure theory was developed in the late 19th and early 20th centuries. Applications of measures include the Lebesgue integral, Kolmogorov's axiomatisation of probability theory, and ergodic theory. Knot theory greatly expanded. Quantum mechanics led to the development of functional analysis. Other new areas include Laurent Schwartz's distribution theory, fixed point theory, singularity theory and René Thom's catastrophe theory, model theory, and Mandelbrot's fractals. Lie theory with its Lie groups and Lie algebras became one of the major areas of study.

Non-standard analysis, introduced by Abraham Robinson, rehabilitated the infinitesimal approach to calculus, which had fallen into disrepute in favour of the theory of limits, by extending the field of real numbers to the Hyperreal numbers which include infinitesimal and infinite quantities. An even larger number system, the surreal numbers were discovered by John Horton Conway in connection with combinatorial games.

The development and continual improvement of computers, at first mechanical analog machines and then digital electronic machines, allowed industry to deal with larger and larger amounts of data to facilitate mass production and distribution and communication, and new areas of mathematics were developed to deal with this: Alan Turing's computability theory; complexity theory; Derrick Henry Lehmer's use of ENIAC to further number theory and the Lucas-Lehmer test; Rózsa Péter's recursive function theory; Claude Shannon's information theory; signal processing; data analysis; optimization and other areas of operations research. In the preceding centuries much mathematical focus was on calculus and continuous functions, but the rise of computing and communication networks led to an increasing importance of discrete concepts and the expansion of combinatorics including graph theory. The speed and data processing abilities of computers also enabled the handling of mathematical problems that were too time-consuming to deal with by pencil and paper calculations, leading to areas such as numerical analysis and symbolic computation. Some of the most important methods and algorithms of the 20th century are: the simplex algorithm, the Fast Fourier Transform, error-correcting codes, the Kalman filter from control theory and the RSA algorithm of public-key cryptography.

At the same time, deep insights were made about the limitations to mathematics. In 1929 and 1930, it was proved the truth or falsity of all statements formulated about the natural numbers plus one of addition and multiplication, was decidable, i.e. could be determined by some algorithm. In 1931, Kurt Gödel found that this was not the case for the natural numbers plus both addition and multiplication; this system, known as Peano arithmetic, was in fact incompletable. (Peano arithmetic is adequate for a good deal of number theory, including the notion of prime number.) A consequence of Gödel's two incompleteness theorems is that in any mathematical system that includes Peano arithmetic (including all of analysis and geometry), truth necessarily outruns proof, i.e. there are true statements that cannot be proved within the system. Hence mathematics cannot be reduced to mathematical logic, and David Hilbert's dream of making all of mathematics complete and consistent needed to be reformulated.

One of the more colorful figures in 20th-century mathematics was Srinivasa Aiyangar Ramanujan (1887–1920), an Indian autodidact who conjectured or proved over 3000 theorems, including properties of highly composite numbers, the partition function and its asymptotics, and mock theta functions. He also made major investigations in the areas of gamma functions, modular forms, divergent series, hypergeometric series and prime number theory.

Paul Erdõs published more papers than any other mathematician in history, working with hundreds of collaborators. Mathematicians have a game equivalent to the Kevin Bacon Game, which leads to the Erdõs number of a mathematician. This describes the "collaborative distance" between a person and Paul Erdõs, as measured by joint authorship of mathematical papers.

Emmy Noether has been described by many as the most important woman in the history of mathematics. She studied the theories of rings, fields, and algebras.

As in most areas of study, the explosion of knowledge in the scientific age has led to specialization: by the end of the century there were hundreds of specialized areas in mathematics and the Mathematics Subject Classification was dozens of pages long. More and more mathematical journals were published and, by the end of the century, the development of the World Wide Web led to online publishing.

21st century

See also: List of unsolved problems in mathematics § Problems solved since 1995

In 2000, the Clay Mathematics Institute announced the seven Millennium Prize Problems, and in 2003 the Poincaré conjecture was solved by Grigori Perelman (who declined to accept an award, as he was critical of the mathematics establishment).

Most mathematical journals now have online versions as well as print versions, and many online-only journals are launched. There is an increasing drive towards open access publishing, first popularized by the arXiv.

Future

There are many observable trends in mathematics, the most notable being that the subject is growing ever larger, computers are ever more important and powerful, the application of mathematics to bioinformatics is rapidly expanding, and the volume of data being produced by science and industry, facilitated by computers, is explosively expanding.

Mathematics (from Greek ìÜèçìá máthçma, "knowledge, study, learning") includes the study of such topics as quantity, structure,space, and change.

Mathematicians seek and use patterns to formulate new conjectures; they resolve the truth or falsity of conjectures by mathematical proof. When mathematical structures are good models of real phenomena, then mathematical reasoning can provide insight or predictions about nature. Through the use of abstraction and logic, mathematics developed from counting, calculation, measurement, and the systematic study of the shapes and motions of physical objects. Practical mathematics has been a human activity from as far back as written recordsexist. The research required to solve mathematical problems can take years or even centuries of sustained inquiry.

Rigorous arguments first appeared in Greek mathematics, most notably in Euclid's Elements. Since the pioneering work of Giuseppe Peano(1858–1932), David Hilbert (1862–1943), and others on axiomatic systems in the late 19th century, it has become customary to view mathematical research as establishing truth by rigorous deduction from appropriately chosen axioms and definitions. Mathematics developed at a relatively slow pace until the Renaissance, when mathematical innovations interacting with new scientific discoveries led to a rapid increase in the rate of mathematical discovery that has continued to the present day.

Mathematics is essential in many fields, including natural science, engineering, medicine, finance, and the social sciences. Applied mathematics has led to entirely new mathematical disciplines, such as statistics and game theory. Mathematicians engage in pure mathematics (mathematics for its own sake) without having any application in mind, but practical applications for what began as pure mathematics are often discovered later.

History

The Babylonian mathematical tablet Plimpton 322, dated to 1800 BC.

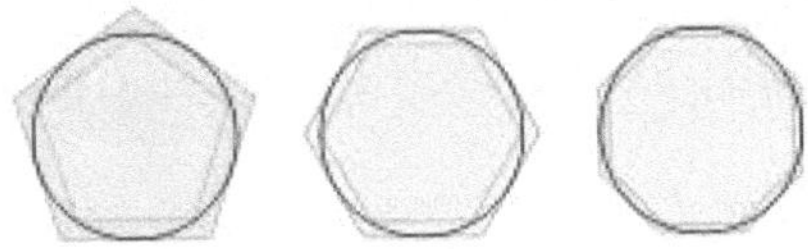

Archimedes used the method of exhaustion to approximate the value of pi.

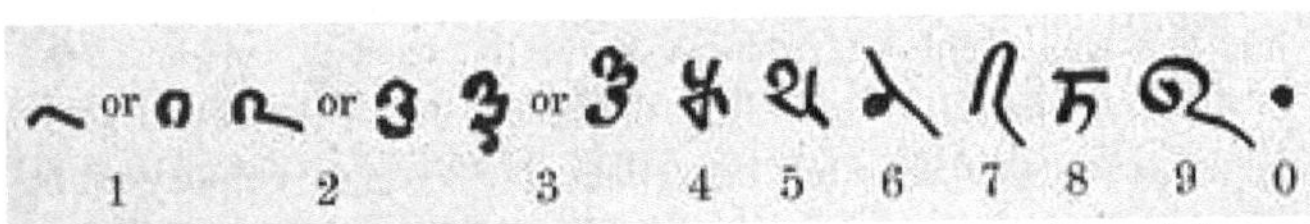

The numerals used in the Bakhshali manuscript, dated between the 2nd century BCE and the 2nd century CE.

The history of mathematics can be seen as an ever-increasing series of abstractions. The first abstraction, which is shared by many animals, was probably that of numbers: the realization that a collection of two apples and a collection of two oranges (for example) have something in common, namely quantity of their members.

As evidenced by tallies found on bone, in addition to recognizing how to count physical objects, prehistoric peoples may have also recognized how to count abstract quantities, like time – days, seasons, years.

Evidence for more complex mathematics does not appear until around 3000 BC, when the Babylonians and Egyptians began using arithmetic, algebra and geometry for taxation and other financial calculations, for building and construction, and for astronomy. The most ancient mathematical texts from Mesopotamia and Egypt are from 2000–1800 BC. Many early texts mention Pythagorean triples and so, by inference, the Pythagorean theorem seems to be the most ancient and widespread mathematical development after basic arithmetic and geometry. It is in Babylonian mathematics that elementary arithmetic (addition, subtraction, multiplication and division) first appear in the archaeological record. The Babylonians also possessed a place-value system, and used a sexagesimal numeral system, still in use today for measuring angles and time.

Beginning in the 6th century BC with the Pythagoreans, the Ancient Greeks began a systematic study of mathematics as a subject in its own right with Greek mathematics. Around 300 BC, Euclid introduced the axiomatic method still used in mathematics today, consisting of definition, axiom, theorem, and proof. His textbook Elements is widely considered the most successful and influential textbook of all time.The greatest mathematician of antiquity is often held to be Archimedes (c. 287–212 BC) of Syracuse. He developed formulas for calculating the surface area and volume of solids of revolution and used the method of exhaustion to calculate the areaunder the arc of a parabola with the summation of an infinite series, in a manner not too dissimilar from modern calculus. Other notable achievements of Greek mathematics are conic sections (Apollonius of Perga, 3rd century BC), trigonometry (Hipparchus of Nicaea (2nd century BC), and the beginnings of algebra (Diophantus, 3rd century AD).

The Hindu–Arabic numeral system and the rules for the use of its operations, in use throughout the world today, evolved over the course of the first millennium AD in India and were transmitted to the Western world via Islamic mathematics. Other notable developments of Indian mathematics include the modern definition of sine and cosine, and an early form of infinite series.

A PAGE FROM AL-KHWÂRIZMÎ'S ALGEBRA

During the Golden Age of Islam, especially during the 9th and 10th centuries, mathematics saw many important innovations building on Greek mathematics. The most notable achievement of Islamic mathematics was the development of algebra. Other notable achievements of the Islamic period are advances in spherical trigonometry and the addition of the decimal point to the Arabic numeral system. Many notable mathematicians from this period were Persian, such as Al-Khwarismi, Omar Khayyam and Sharaf al-Dîn al-lûsî.

During the early modern period, mathematics began to develop at an accelerating pace in Western Europe. The development of calculus by Newton and Leibniz in the 17th century revolutionized mathematics. Leonhard Euler was the most notable mathematician of the 18th century, contributing numerous theorems and discoveries. Perhaps the foremost mathematician of the 19th century was the German mathematician Carl Friedrich Gauss, who made numerous contributions to fields such as algebra, analysis, differential geometry, matrix theory,number theory, and statistics. In the early 20th century, Kurt Gödel transformed mathematics by publishing his incompleteness theorems, which show that any axiomatic system that is consistent will contain unprovable propositions.

Mathematics has since been greatly extended, and there has been a fruitful interaction between mathematics and science, to the benefit of both. Mathematical discoveries continue to be made today. According to Mikhail B. Sevryuk, in the January 2006 issue of the Bulletin of the American Mathematical Society, "The number of papers and books included in the Mathematical Reviews database since 1940 (the first year of operation of MR) is now more than 1.9 million, and more than 75 thousand items are added to the database each year. The overwhelming majority of works in this ocean contain new mathematical theorems and their proofs."

Etymology

The word mathematics comes from Ancient Greek ìÜèçìá (máthçma), meaning "that which is learnt", "what one gets to know", hence also "study" and "science". The word for "mathematics" came to have the narrower and more technical meaning "mathematical study" even in Classical times. Its adjective is ìáèçìáôéêüò (mathçmatikós), meaning "related to learning" or "studious", which likewise further came to mean "mathematical". In particular, ìáèçìáôéêÞ ôÝ÷íç (mathçmatik tékhnç), Latin: ars mathematica, meant "the mathematical art".

Similarly, one of the two main schools of thought in Pythagoreanism was known as the mathçmatikoi (ìáèçìáôéêïß)—which at the time meant "teachers" rather than "mathematicians" in the modern sense.

In Latin, and in English until around 1700, the term mathematics more commonly meant "astrology" (or sometimes "astronomy") rather than "mathematics"; the meaning gradually changed to its present one from about 1500 to 1800. This has resulted in several mistranslations. For example, Saint Augustine's warning that Christians should beware of mathematici, meaning astrologers, is sometimes mistranslated as a condemnation of mathematicians.

The apparent plural form in English, like the French plural form les mathématiques (and the less commonly used singular derivative la mathématique), goes back to the Latin neuter plural mathematica (Cicero), based on the Greek plural ôp ìáèçìáôéêÜ (ta mathçmatiká), used by Aristotle (384–322 BC), and meaning roughly "all things mathematical"; although it is plausible that English borrowed only the adjective mathematic(al) and formed the noun mathematics anew, after the pattern of physics and metaphysics, which were inherited from Greek. In English, the noun mathematics takes a singular verb. It is often shortened to maths or, in North America, math.

Definitions of mathematics

Mathematics has no generally accepted definition. Aristotle defined mathematics as "the science of quantity", and this definition prevailed until the 18th century. Galileo Galilei (1564–1642) said, "The universe cannot be read until we have learned the language and become familiar with the characters in which it is written. It is written in mathematical language, and the letters are triangles, circles and other geometrical figures, without which means it is humanly impossible to comprehend a single word. Without these, one is wandering about in a dark labyrinth." Carl Friedrich Gauss (1777–1855) referred to mathematics as "the Queen of the Sciences". Benjamin Peirce (1809–1880) called mathematics "the science that draws necessary conclusions". David Hilbert said of mathematics: "We are not speaking here of arbitrariness in any sense. Mathematics is not like a game whose tasks are determined by arbitrarily stipulated rules. Rather, it is a conceptual system possessing internal necessity that can only be so and by no means otherwise." Albert Einstein (1879–1955) stated that "as far as the laws of mathematics refer to reality, they are not certain; and as far as they are certain, they do not refer to reality."

Starting in the 19th century, when the study of mathematics increased in rigor and began to address abstract topics such as group theory and projective geometry, which have no clear-cut relation to quantity and measurement, mathematicians and philosophers began to propose a variety of new definitions. Some of these definitions emphasize the deductive character of much of mathematics, some emphasize its abstractness, some emphasize certain topics within mathematics. Today, no consensus on the definition of mathematics prevails, even among professionals. There is not even consensus on whether mathematics is an art or a science. A great many professional mathematicians take no interest in a definition of mathematics, or consider it undefinable. Some just say, "Mathematics is what mathematicians do."

Three leading types of definition of mathematics are called logicist, intuitionist, and formalist, each reflecting a different philosophical school of thought. All have severe problems, none has widespread acceptance, and no reconciliation seems possible.

An early definition of mathematics in terms of logic was Benjamin Peirce's "the science that draws necessary conclusions" (1870). In the Principia Mathematica, Bertrand Russell and Alfred North Whitehead advanced the philosophical program known as logicism, and attempted to prove that all mathematical concepts, statements, and principles can be defined and proved entirely in terms of symbolic logic. A logicist definition of mathematics is Russell's "All Mathematics is Symbolic Logic" (1903).

Intuitionist definitions, developing from the philosophy of mathematician L. E. J. Brouwer, identify mathematics with certain mental phenomena. An example of an intuitionist definition is "Mathematics is the mental activity which consists in carrying out constructs one after the other." A peculiarity of intuitionism is that it rejects some mathematical ideas considered valid according to other definitions. In particular, while other philosophies of mathematics allow objects that can be proved to exist even though they cannot be constructed, intuitionism allows only mathematical objects that one can actually construct.

Formalist definitions identify mathematics with its symbols and the rules for operating on them. Haskell Curry defined mathematics simply as "the science of formal systems". A formal system is a set of symbols, or tokens, and some rules telling how the tokens may be combined into formulas. In formal systems, the word axiom has a special meaning, different from the ordinary meaning of "a self-evident truth". In formal systems, an axiom is a combination of tokens that is included in a given formal system without needing to be derived using the rules of the system.

Mathematics as science

The German mathematician Carl Friedrich Gauss referred to mathematics as "the Queen of the Sciences". More recently, Marcus du Sautoy has called mathematics "the Queen of Science ... the main driving force behind scientific discovery". In the original Latin Regina Scientiarum, as well as in German Königin der Wissenschaften, the word corresponding to science means a "field of knowledge", and this was the original meaning of "science" in English, also; mathematics is in this sense a field of knowledge. The specialization restricting the meaning of "science" to natural sciencefollows the rise of Baconian science, which contrasted "natural science" to scholasticism, the Aristotelean method of inquiring from first principles. The role of empirical experimentation and observation is negligible in mathematics, compared to natural sciences such as biology, chemistry, or physics. Albert Einstein stated that "as far as the laws of mathematics refer to reality, they are not certain; and as far as they are certain, they do not refer to reality."

Many philosophers believe that mathematics is not experimentally falsifiable, and thus not a science according to the definition of Karl Popper.However, in the 1930s Gödel's incompleteness theorems convinced many mathematicians that mathematics cannot be reduced to logic alone, and Karl Popper concluded that "most mathematical theories are, like those of physics and biology, hypothetico-deductive: pure mathematics therefore turns out to be much closer to the natural sciences whose hypotheses are conjectures, than it seemed even recently." Other thinkers, notably Imre Lakatos, have applied a version of falsificationism to mathematics itself.

An alternative view is that certain scientific fields (such as theoretical physics) are mathematics with axioms that are intended to correspond to reality. Mathematics shares much in common with many fields in the physical sciences, notably the exploration of the logical consequences of assumptions. Intuition and experimentation also play a role in the formulation of conjectures in both mathematics and the (other) sciences. Experimental mathematics continues to grow in importance within mathematics, and computation and simulation are playing an increasing role in both the sciences and mathematics.

The opinions of mathematicians on this matter are varied. Many mathematicians feel that to call their area a science is to downplay the importance of its aesthetic side, and its history in the traditional seven liberal arts; others feel that to ignore its connection to the sciences is to turn a blind eye to the fact that the interface between mathematics and its applications in science and engineering has driven much development in mathematics. One way this difference of viewpoint plays out is in the philosophical debate as to whether mathematics is created (as in art) or discovered (as in science). It is common to see universities divided into sections that include a division of Science and Mathematics, indicating that the fields are seen as being allied but that they do not coincide. In practice, mathematicians are typically grouped with scientists at the gross level but separated at finer levels. This is one of many issues considered in the philosophy of mathematics.

Inspiration, pure and applied mathematics, and aesthetics

Mathematics arises from many different kinds of problems. At first these were found in commerce, land measurement, architecture and later astronomy; today, all sciences suggest problems studied by mathematicians, and many problems arise within mathematics itself. For example, the physicist Richard Feynman invented the path integral formulation of quantum mechanics using a combination of mathematical reasoning and physical insight, and today's string theory, a still-developing scientific theory which attempts to unify the four fundamental forces of nature, continues to inspire new mathematics.

Some mathematics is relevant only in the area that inspired it, and is applied to solve further problems in that area. But often mathematics inspired by one area proves useful in many areas, and joins the general stock of mathematical concepts. A distinction is often made between pure mathematics and applied mathematics. However pure mathematics topics often turn out to have applications, e.g. number theory in cryptography. This remarkable fact, that even the "purest" mathematics often turns out to have practical applications, is what Eugene Wigner has called "the unreasonable effectiveness of mathematics". As in most areas of study, the explosion of knowledge in the scientific age has led to specialization: there are now hundreds of specialized areas in mathematics and the latest Mathematics Subject Classification runs to 46 pages. Several areas of applied mathematics have merged with related traditions outside of mathematics and become disciplines in their own right, including statistics, operations research, and computer science.

For those who are mathematically inclined, there is often a definite aesthetic aspect to much of mathematics. Many mathematicians talk about the elegance of mathematics, its intrinsic aesthetics and inner beauty. Simplicity and generality are valued. There is beauty in a simple and elegant proof, such as Euclid's proof that there are infinitely many prime numbers, and in an elegant numerical method that speeds calculation, such as the fast Fourier transform. G. H. Hardy in A Mathematician's Apology expressed the belief that these aesthetic considerations are, in themselves, sufficient to justify the study of pure mathematics. He identified criteria such as significance, unexpectedness, inevitability, and economy as factors that contribute to a mathematical aesthetic. Mathematical research often seeks critical features of a mathematical object. A theorem expressed as a characterization of the object by these features is the prize. Examples of particularly succinct and revelatory mathematical arguments has been published in Proofs from THE BOOK.

The popularity of recreational mathematics is another sign of the pleasure many find in solving mathematical questions. And at the other social extreme, philosophers continue to find problems in philosophy of mathematics, such as the nature of mathematical proof.

Notation, Language, and Rigor

Most of the mathematical notation in use today was not invented until the 16th century. Before that, mathematics was written out in words, limiting mathematical discovery. Euler (1707–1783) was responsible for many of the notations in use today. Modern notation makes mathematics much easier for the professional, but beginners often find it daunting. According to Barbara Oakley, this can be attributed to the fact that mathematical ideas are both more abstract and more encrypted than those of natural language. Unlike natural language, where people can often equate a word (such as cow) with the physical object it corresponds to, mathematical symbols are abstract, lacking any physical analog. Mathematical symbols are also more highly encrypted than regular words, meaning a single symbol can encode a number of different operations or ideas.

Mathematical language can be difficult to understand for beginners because even common terms, such as or and only, have a more precise meaning than they have in everyday speech, and other terms such as open and field refer to specific mathematical ideas, not covered by their laymen's meanings. Mathematical language also includes many technical terms such as homeomorphism and integrable that have no meaning outside of mathematics. Additionally, shorthand phrases such as iff for "if and only if" belong to mathematical jargon. There is a reason for special notation and technical vocabulary: mathematics requires more precision than everyday speech. Mathematicians refer to this precision of language and logic as "rigor".

Mathematical proof is fundamentally a matter of rigor. Mathematicians want their theorems to follow from axioms by means of systematic reasoning. This is to avoid mistaken "theorems", based on fallible intuitions, of which many instances have occurred in the history of the subject. The level of rigor expected in mathematics has varied over time: the Greeks expected detailed arguments, but at the time of Isaac Newton the methods employed were less rigorous. Problems inherent in the definitions used by Newton would lead to a resurgence of careful analysis and formal proof in the 19th century. Misunderstanding the rigor is a cause for some of the common misconceptions of mathematics. Today, mathematicians continue to argue among themselves about computer-assisted proofs. Since large computations are hard to verify, such proofs may not be sufficiently rigorous.

Axioms in traditional thought were "self-evident truths", but that conception is problematic. At a formal level, an axiom is just a string of symbols, which has an intrinsic meaning only in the context of all derivable formulas of an axiomatic system. It was the goal of Hilbert's program to put all of mathematics on a firm axiomatic basis, but according to Gödel's incompleteness theorem every (sufficiently powerful) axiomatic system has undecidable formulas; and so a final axiomatization of mathematics is impossible. Nonetheless mathematics is often imagined to be (as far as its formal content) nothing but set theory in some axiomatization, in the sense that every mathematical statement or proof could be cast into formulas within set theory.

Fields of mathematics

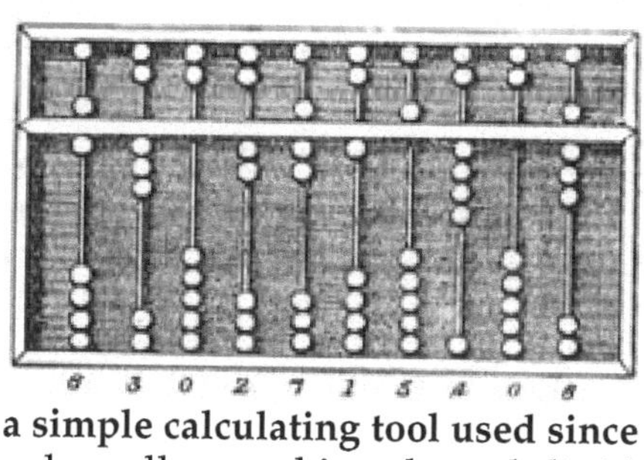

The abacus is a simple calculating tool used since ancient times.

Mathematics can, broadly speaking, be subdivided into the study of quantity, structure, space, and change (i.e. arithmetic, algebra, geometry, and analysis). In addition to these main concerns, there are also subdivisions dedicated to exploring links from the heart of mathematics to other fields: to logic, to set theory (foundations), to the empirical mathematics of the various sciences (applied mathematics), and more recently to the rigorous study of uncertainty. While some areas might seem unrelated, the Langlands program has found connections between areas previously thought unconnected, such as Galois groups, Riemann surfaces and number theory.

Foundations and philosophy

In order to clarify the foundations of mathematics, the fields of mathematical logic and set theory were developed. Mathematical logic includes the mathematical study of logic and the applications of formal logic to other areas of mathematics; set theory is the branch of mathematics that studies sets or collections of objects. Category theory, which deals in an abstract way with mathematical structures and relationships between them, is still in development. The phrase "crisis of foundations" describes the search for a rigorous foundation for mathematics that took place from approximately 1900 to 1930. Some disagreement about the foundations of mathematics continues to the present day. The crisis of foundations was stimulated by a number of controversies at the time, including the controversy over Cantor's set theory and the Brouwer–Hilbert controversy.

Mathematical logic is concerned with setting mathematics within a rigorous axiomatic framework, and studying the implications of such a framework. As such, it is home to Gödel's incompleteness theorems which (informally) imply that any effective formal system that contains basic arithmetic, if sound (meaning that all theorems that can be proved are true), is necessarily incomplete (meaning that there are true theorems which cannot be proved in that system). Whatever finite collection of number-theoretical axioms is taken as a foundation, Gödel showed how to construct a formal statement that is a true number-theoretical fact, but which does not follow from those axioms. Therefore, no formal system is a complete axiomatization of full number theory. Modern logic is divided into recursion theory, model theory, and proof theory, and is closely linked to theoretical computer science, as well as to category theory. In the context of recursion theory, the impossibility of a full axiomatization of number theory can also be formally demonstrated as a consequence of the MRDP theorem.

Theoretical computer science includes computability theory, computational complexity theory, and information theory. Computability theory examines the limitations of various theoretical models of the computer, including the most well-known model – the Turing machine. Complexity theory is the study of tractability by computer; some problems, although theoretically solvable by computer, are so expensive in terms of time or space that solving them is likely to remain practically unfeasible, even with the rapid advancement of computer hardware. A famous problem is the "P = NP?" problem, one of the Millennium Prize Problems. Finally, information theory is concerned with the amount of data that can be stored on a given medium, and hence deals with concepts such as compression and entropy.

	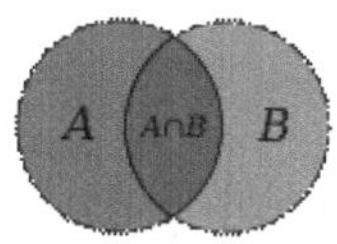	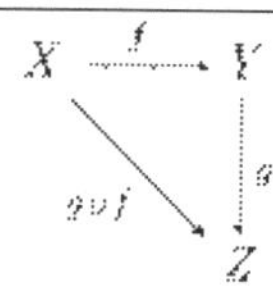	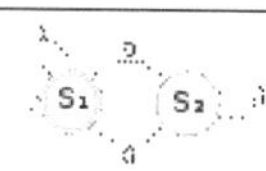
Mathematical logic	Set theory	Category theory	Theory of computation

Pure mathematics

Quantity

The study of quantity starts with numbers, first the familiar natural numbers and integers ("whole numbers") and arithmetical operations on them, which are characterized in arithmetic. The deeper properties of integers are studied in number theory, from which come such popular results as Fermat's Last Theorem. The twin prime conjecture and Goldbach's conjecture are two unsolved problems in number theory.

As the number system is further developed, the integers are recognized as a subset of the rational numbers ("fractions"). These, in turn, are contained within the real numbers, which are used to represent continuous quantities. Real numbers are generalized to complex numbers. These are the first steps of a hierarchy of numbers that goes on to include quaternions and octonions. Consideration of the natural numbers also leads to the transfinite numbers, which formalize the concept of "infinity". According to the fundamental theorem of algebra all solutions of equations in one unknown with complex coefficients are complex numbers, regardless of degree. Another area of study is the size of sets, which is described with the cardinal numbers. These include the aleph numbers, which allow meaningful comparison of the size of infinitely large sets.

Structure

Many mathematical objects, such as sets of numbers and functions, exhibit internal structure as a consequence of operations or relations that are defined on the set. Mathematics then studies properties of those sets that can be expressed in terms of that structure; for instance number theory studies properties of the set of integers that can be expressed in terms of arithmetic operations. Moreover, it frequently happens that different such structured sets (or structures) exhibit similar properties, which makes it possible, by a further step of abstraction, to state axioms for a class of structures, and then study at once the whole class of structures satisfying these axioms. Thus one can study groups, rings, fields and other abstract systems; together such studies (for structures defined by algebraic operations) constitute the domain of abstract algebra.

By its great generality, abstract algebra can often be applied to seemingly unrelated problems; for instance a number of ancient problems concerning compass and straightedge constructions were finally solved using Galois theory, which involves field theory and group theory. Another example of an algebraic theory is linear algebra, which is the general study of vector spaces, whose elements called vectors have both quantity and direction, and can be used to model (relations between) points in space. This is one example of the phenomenon that the originally unrelated areas of geometry and algebra have very strong interactions in modern mathematics. Combinatorics studies ways of enumerating the number of objects that fit a given structure.

	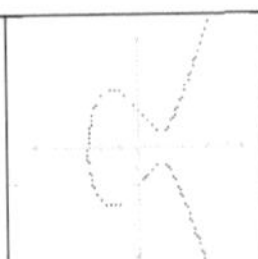		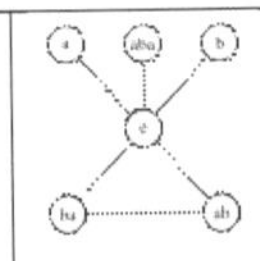	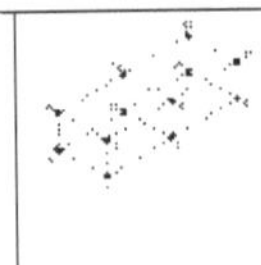	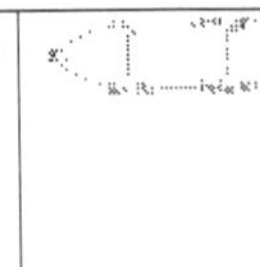
Combinatorics	Number theory	Group theory	Graph theory	Order theory	Algebra

Space

The study of space originates with geometry – in particular, Euclidean geometry, which combines space and numbers, and encompasses the well-known Pythagorean theorem. Trigonometry is the branch of mathematics that deals with relationships between the sides and the angles of triangles and with the trigonometric functions. The modern study of space generalizes these ideas to include higher-dimensional geometry, non-Euclidean geometries (which play a central role in general relativity) and topology. Quantity and space both play a role in analytic geometry, differential geometry, and algebraic geometry. Convex and discrete geometry were developed to solve problems in number theory and functional analysis but now are pursued with an eye on applications in optimization and computer science. Within differential geometry are the concepts of fiber bundles and calculus on manifolds, in particular, vector and tensor calculus. Within algebraic geometry is the description of geometric objects as solution sets of polynomial equations, combining the concepts of quantity and space, and also the study of topological groups, which combine structure and space. Lie groups are used to study space, structure, and change. Topology in all its many ramifications may have been the greatest growth area in 20th-century mathematics; it includes point-set topology, set-theoretic topology, algebraic topology and differential topology. In particular, instances of modern-day topology are metrizability theory, axiomatic set theory, homotopy theory, and Morse theory. Topology also includes the now solved Poincaré conjecture, and the still unsolved areas of the Hodge conjecture. Other results in geometry and topology, including the four color theorem and Kepler conjecture, have been proved only with the help of computers.

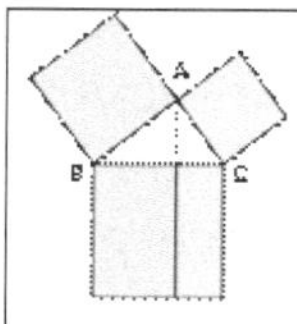	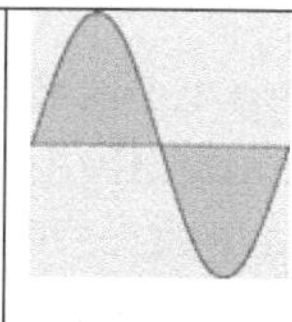	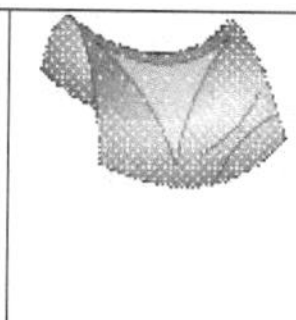	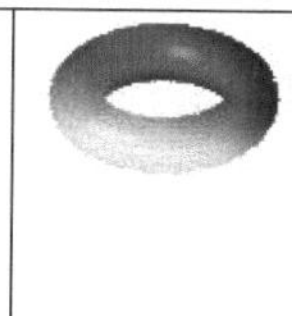	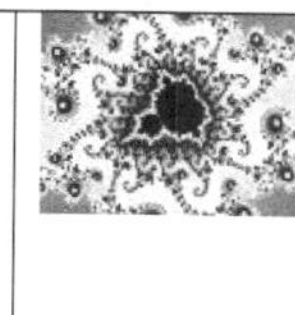	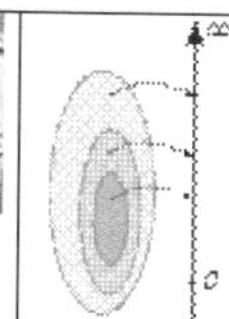
Geometry	Trigonometry	Differential geometry	Topology	Fractal geometry	Measure theory

Change

Understanding and describing change is a common theme in the natural sciences, and calculus was developed as a powerful tool to investigate it. Functions arise here, as a central concept describing a changing quantity. The rigorous study of real numbers and functions of a real variable is known as real analysis, with complex analysis the equivalent field for the complex numbers. Functional analysis focuses attention on (typically infinite-dimensional) spaces of functions. One of many applications of functional analysis is quantum mechanics. Many problems lead naturally to relationships between a quantity and its rate of change, and these are studied as differential equations. Many phenomena in nature can be described by dynamical systems; chaos theory makes precise the ways in which many of these systems exhibit unpredictable yet still deterministic behavior.

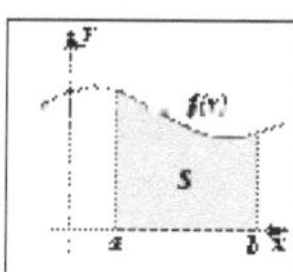	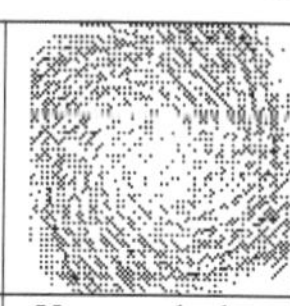	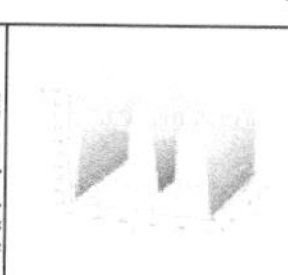	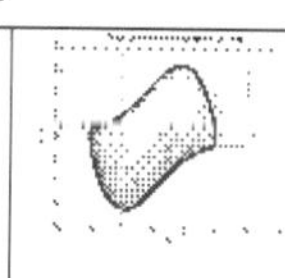	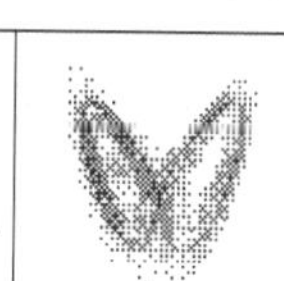	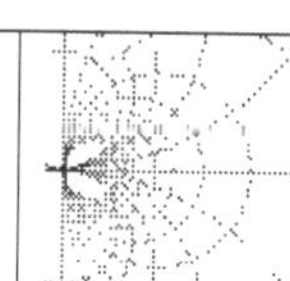
Calculus	Vector calculus	Differential equations	Dynamical systems	Chaos theory	Complex analysis

Applied mathematics

Applied mathematics concerns itself with mathematical methods that are typically used in science, engineering, business, and industry. Thus, "applied mathematics" is a mathematical science with specialized knowledge. The term applied mathematics also describes the professional specialty in which mathematicians work on practical problems; as a profession focused on practical problems, applied mathematics focuses on the "formulation, study, and use of mathematical models" in science, engineering, and other areas of mathematical practice.

In the past, practical applications have motivated the development of mathematical theories, which then became the subject of study in pure mathematics, where mathematics is developed primarily for its own sake. Thus, the activity of applied mathematics is vitally connected with research in pure mathematics.

Statistics and other decision sciences

Applied mathematics has significant overlap with the discipline of statistics, whose theory is formulated mathematically, especially with probability theory. Statisticians (working as part of a research project) "create data that makes sense" with random sampling and with randomized experiments; the design of a statistical sample or experiment specifies the analysis of the data (before the data be available). When reconsidering data from experiments and samples or when analyzing data from observational studies, statisticians "make sense of the data" using the art of modelling and the theory of inference – with model selection and estimation; the estimated models and consequential predictions should be testedon new data.

Statistical theory studies decision problems such as minimizing the risk (expected loss) of a statistical action, such as using a procedure in, for example, parameter estimation, hypothesis testing, and selecting the best. In these traditional areas of mathematical statistics, a statistical-decision problem is formulated by minimizing an objective function, like expected loss or cost, under specific constraints: For example, designing a survey often involves minimizing the cost of estimating a population mean with a given level of confidence. Because of its use of optimization, the mathematical theory of statistics shares concerns with other decision sciences, such as operations research, control theory, and mathematical economics.

Computational mathematics

Computational mathematics proposes and studies methods for solving mathematical problems that are typically too large for human numerical capacity. Numerical analysis studies methods for problems in analysis using functional analysis and approximation theory; numerical analysis includes the study of approximation and discretization broadly with special concern for rounding errors. Numerical analysis and, more broadly, scientific computing also study non-analytic topics of mathematical science, especially algorithmic matrix and graph theory. Other areas of computational mathematics include computer algebra and symbolic computation.

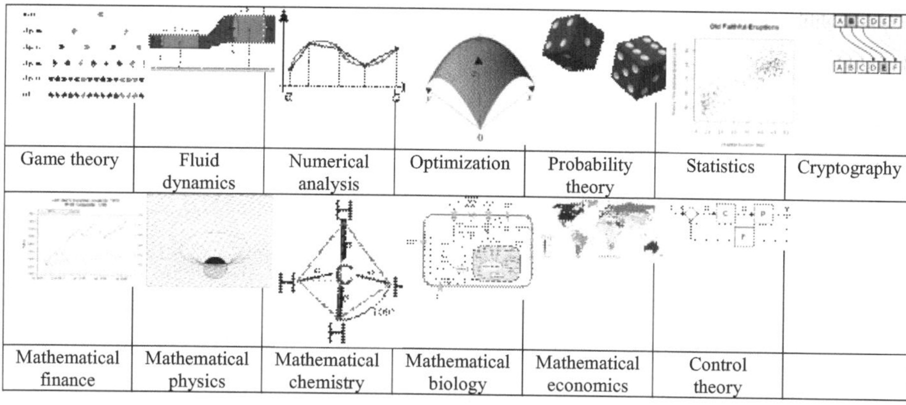

Mathematical awards

Arguably the most prestigious award in mathematics is the Fields Medal, established in 1936 and awarded every four years (except around World War II) to as many as four individuals. The Fields Medal is often considered a mathematical equivalent to the Nobel Prize.

The Wolf Prize in Mathematics, instituted in 1978, recognizes lifetime achievement, and another major international award, the Abel Prize, was instituted in 2003. The Chern Medalwas introduced in 2010 to recognize lifetime achievement. These accolades are awarded in recognition of a particular body of work, which may be innovational, or provide a solution to an outstanding problem in an established field.

A famous list of 23 open problems, called "Hilbert's problems", was compiled in 1900 by German mathematician David Hilbert. This list achieved great celebrity among mathematicians, and at least nine of the problems have now been solved. A new list of seven important problems, titled the "Millennium Prize Problems", was published in 2000. Only one of them, the Riemann hypothesis, duplicates one of Hilbert's problems. A solution to any of these problems carries a $1 million reward.

2

Areas of Mathematics

Mathematics encompasses a growing variety and depth of subjects over history, and comprehension requires a system to categorize and organize the many subjects into more general areas of mathematics. A number of different classification schemes have arisen, and though they share some similarities, there are differences due in part to the different purposes they serve. In addition, as mathematics continues to be developed, these classification schemes must change as well to account for newly created areas or newly discovered links between different areas. Classification is made more difficult by some subjects, often the most active, which straddle the boundary between different areas.

A traditional division of mathematics is into pure mathematics, mathematics studied for its intrinsic interest, and applied mathematics, mathematics which can be directly applied to real world problems. This division is not always clear and many subjects have been developed as pure mathematics to find unexpected applications later on. Broad divisions, such as discrete mathematics and computational mathematics, have emerged more recently.

An ideal system of classification permits adding new areas into the organization of previous knowledge, and fitting surprising discoveries and unexpected interactions into the outline. For example, the Langlands program has found unexpected connections between areas previously thought unconnected, at least Galois groups, Riemann surfaces and number theory.

CLASSIFICATION SYSTEMS

The Mathematics Subject Classification (MSC) is produced by the staff of the review databases Mathematical Reviews and Zentralblatt MATH. Many mathematics journals ask authors to label their papers with MSC subject codes. The MSC divides mathematics into over 60 areas, with further subdivisions within each area.

- In the Library of Congress Classification, mathematics is assigned the subclass QA within the class Q (Science). The LCC defines broad divisions, and individual subjects are assigned specific numerical values.
- The Dewey Decimal Classification assigns mathematics to division 510, with subdivisions for Algebra & number theory, Arithmetic, Topology, Analysis, Geometry, Numerical analysis, and Probabilities & applied mathematics.
- The Categories within Mathematics list is used by the Arxiv for categorizing preprints. It differs from MSC; for example, it includes things like quantum algebra.
- The IMU uses its programme structure for organizing the lectures at its four-yearly ICM. One of its top-level sections that MSC doesn't have is Lie theory.
- The ACM Computing Classification System includes a couple of mathematical categories F. Theory of Computation and G. Mathematics of Computing.
- MathOverflow has a tag system.
- Mathematics book publishers such as Springer (subdisciplines), Cambridge (Browse Mathematics and statistics) and the AMS (subject area) use their own subject lists on their websites to enable customers to browse books or filter searches by subdiscipline, including topics such as mathematical biology and mathematical finance as top-level headings.
- Schools and other educational bodies have syllabuses.
- Research institutes and university mathematics departments often have sub-departments or study groups. e.g. SIAM has activity groups for its members.
- Wikipedia uses a Category: Mathematics system on its articles, and also has a list of mathematics lists.

MAJOR DIVISIONS OF MATHEMATICS

Foundations

From magic squares to the Mandelbrot set, numbers have been a source of amusement and delight for millions of people throughout the ages. Many important branches of "serious" mathematics have their roots in what was once a mere puzzle and/or game.

History and biography

The history of mathematics is inextricably intertwined with the subject itself. This is perfectly natural: mathematics has an internal organic structure, deriving new theorems from those that have come before. As each new generation of mathematicians builds upon the achievements of our ancestors, the subject itself expands and grows new layers, like an onion.

Mathematical logic and foundations, including set theory

Mathematicians have always worked with logic and symbols, but for centuries the underlying laws of logic were taken for granted, and never expressed symbolically. Mathematical logic, also known as symbolic logic, was developed when people finally realized that the tools of mathematics can be used to study the structure of logic itself. Areas of research in this field have expanded rapidly, and are usually subdivided into several distinct departments.

Model theory

Model theory studies mathematical structures in a general framework. Its main tool is first-order logic.

Set theory

A set can be thought of as a collection of distinct things united by some common feature. Set theory is subdivided into three main areas. Naive set theory is the original set theory developed by mathematicians at the end of the 19th century. Axiomatic set theory is a rigorous axiomatic theory developed in response to the discovery of serious flaws (such as Russell's paradox) in naive set theory. It treats sets as "whatever satisfies the axioms", and the notion of collections of things serves only as motivation for the axioms. Internal set theory is an axiomatic extension of set theory that supports a logically consistent identification of illimited (enormously large) and infinitesimal(unimaginably small) elements within the real numbers. See also List of set theory topics.

Proof theory and constructive mathematics

Proof theory grew out of David Hilbert's ambitious program to formalize all the proofs in mathematics. The most famous result in the field is encapsulated in Gödel's incompleteness theorems. A closely related and now quite popular concept is the idea of Turing machines. Constructivism is the outgrowth of Brouwer's unorthodox view of the nature of logic itself; constructively speaking, mathematicians cannot assert "Either a circle is round, or it is not" until they have actually exhibited a circle and measured its roundness.

Arithmetic

Arithmetic is the study of numbers and the properties of operations between them.

Algebra

The study of structure begins with numbers, first the familiar natural numbers and integers and their arithmetical operations, which are recorded in elementary algebra. The deeper properties of these numbers are studied in number theory. The investigation of methods to solve equations leads to the field of abstract algebra, which, among other things, studies rings and fields, structures that generalize the properties possessed by everyday numbers. Long standing questions about compass and straightedge construction were finally settled by Galois theory. The physically important concept of vectors, generalized to vector spaces, is studied in linear algebra.

ORDER THEORY

For any two distinct real numbers, one must be greater than the other. Order Theory extends this idea to sets in general. It includes notions like lattices and ordered algebraic structures. See also the order theory glossary and the list of order topics.

General algebraic systems

Given a set, different ways of combining or relating members of that set can be defined. If these obey certain rules, then a particular algebraic structure is formed. Universal algebra is the more formal study of these structures and systems.

Number theory

Number theory is traditionally concerned with the properties of integers. More recently, it has come to be concerned with wider classes of problems that have arisen naturally from the study of integers. It can be divided into elementary number theory (where the integers are studied without the aid of techniques from other mathematical fields); analytic number theory (where calculus and complex analysis are used as tools); algebraic number theory (which studies the algebraic numbers - the roots of polynomials with integer coefficients); geometric number theory; combinatorial number theory; transcendental number theory; and computational number theory. See also the list of number theory topics.

Field theory and polynomials

Field theory studies the properties of fields. A field is a mathematical entity for which addition, subtraction, multiplication and division are well-defined. A polynomial is an expression in which constants and variables are combined using only addition, subtraction, and multiplication.

Commutative rings and algebras

In ring theory, a branch of abstract algebra, a commutative ring is a ring in which the multiplication operation obeys the commutative law. This means that if a and b are any elements of the ring, then $a \times b = b \times a$. Commutative algebra is the field of study of commutative rings and their ideals, modules and algebras. It is foundational both for algebraic geometry and for algebraic number theory. The most prominent examples of commutative rings are rings of polynomials.

Analysis

Within the world of mathematics, analysis is the branch that focuses on change: rates of change, accumulated change, and multiple things changing relative to (or independently of) one another.

Modern analysis is a vast and rapidly expanding branch of mathematics that touches almost every other subdivision of the discipline, finding direct and indirect applications in topics as diverse as number theory, cryptography, and abstract algebra. It is also the language of science itself and is used across chemistry, biology, and physics, from astrophysics to X-ray crystallography.

Combinatorics

Combinatorics is the study of finite or discrete collections of objects that satisfy specified criteria. In particular, it is concerned with "counting" the objects in those collections (enumerative combinatorics) and with deciding whether certain "optimal" objects exist (extremal combinatorics). It includes graph theory, used to describe inter-connected objects (a graph in this sense is a network, or collection of connected points). See also the list of combinatorics topics, list of graph theory topics and glossary of graph theory. A combinatorial flavour is present in many parts of problem-solving.

Geometry and topology

Geometry deals with spatial relationships, using fundamental qualities or axioms. Such axioms can be used in conjunction with mathematical definitions for points, straight lines, curves, surfaces, and solids to draw logical conclusions. See also List of geometry topics

Convex geometry and discrete geometry

Includes the study of objects such as polytopes and polyhedra. See also List of convexity topics

Discrete or combinatorial geometry

The study of geometrical objects and properties that are discrete or combinatorial, either by their nature or by their representation. It includes the study of shapes such as the Platonic solids and the notion of tessellation.

Differential geometry

The study of geometry using calculus. It is very closely related to differential topology. Covers such areas as Riemannian geometry, curvature and differential geometry of curves. See also the glossary of differential geometry and topology.

ALGEBRAIC GEOMETRY

Given a polynomial of two real variables, then the points on a plane where that function is zero will form a curve. An algebraic curve extends this notion to polynomials over a fieldin a given number of variables. Algebraic geometry may be viewed as the study of these curves. See also the list of algebraic geometry topics and list of algebraic surfaces.

Topology

Deals with the properties of a figure that do not change when the figure is continuously deformed. The main areas are point set topology (or general topology), algebraic topology, and the topology of manifolds, defined below.

General topology

Also called point set topology. Properties of topological spaces. Includes such notions as open and closed sets, compact spaces, continuous functions, convergence, separation axioms, metric spaces, dimension theory. See also the glossary of general topology and the list of general topology topics.

Algebraic topology

Properties of algebraic objects associated with a topological space and how these algebraic objects capture properties of such spaces. Contains areas like homology theory, cohomology theory, homotopy theory, and homological algebra, some of them examples of functors. Homotopy deals with homotopy groups (including the fundamental group) as well as simplicial complexes and CW complexes (also called cell complexes). See also the list of algebraic topology topics.

Manifolds

A manifold can be thought of as an n-dimensional generalization of a surface in the usual 3-dimensional Euclidean space. The study of manifolds includes differential topology, which looks at the properties of differentiable functions defined over a manifold. See also complex manifolds.

Applied mathematics

Probability and statistics

- Probability theory: The mathematical theory of random phenomena. Probability theory studies random variables and events, which are mathematical abstractions of non-deterministic events or measured quantities. See also Category:probability theory, and the list of probability topics.
- Stochastic processes: An extension of probability theory that studies collections of random variables, such as time series or spatial processes. See also List of stochastic processes topics, and Category:Stochastic processes.
- Statistics: The science of making effective use of numerical data from experiments or from populations of individuals. Statistics includes not only the collection, analysis and interpretation of such data, but also the planning of the collection of data, in terms of the design of surveys and experiments. See also the list of statistical topics.

Computational sciences

Many problems in mathematics cannot in general be solved exactly. Numerical analysis is the study of iterative methods and algorithms for approximately solving problems to a specified error bound. Includes numerical differentiation, numerical integration and numerical methods; c.f. scientific computing. See also List of numerical analysis topics

Computer algebra

This area is also called symbolic computation or algebraic computation. It deals with exact computation, for example with integers of arbitrary size, polynomials or elements of finite fields. It includes also the computation with non numeric mathematical objects like polynomial ideals or series.

Physical sciences

Addresses what happens when a real physical object is subjected to forces. This divides naturally into the study of rigid solids, deformable solids, and fluids, detailed below.

Further information: Continuum mechanics, Elasticity (physics), and Plasticity (physics)

Mechanics of structures

Mechanics of structures is a field of study within applied mechanics that investigates the behavior of structures under mechanical loads, such as bending of a beam, buckling of a column, torsion of a shaft, deflection of a thin shell, and vibration of a bridge.

Further information: Energy principles in structural mechanics, Flexibility method, Direct stiffness method, and Finite element method

Mechanics of deformable solids

Most real-world objects are not point-like nor perfectly rigid. More importantly, objects change shape when subjected to forces. This subject has a very strong overlap with continuum mechanics, which is concerned with continuous matter. It deals with such notions as stress, strain and elasticity.

Further information: Solid state physics, Materials science, Mechanics of materials, Mechanics of solids, Fracture mechanics, Deformation (mechanics), and Deformable bodies

Fluid mechanics

Fluids in this sense includes not just liquids, but flowing gases, and even solids under certain situations. (For example, dry sand can behave like a fluid). It includes such notions as viscosity, turbulent flow and laminar flow (its opposite).

Further information: Fluid dynamics, Mechanics of fluids, Rheology, Electrodynamics, Plasma Physics, Gas dynamics, and Aerodynamics

Particle mechanics

In mathematics, a particle is a point-like, perfectly rigid, solid object. Particle mechanics deals with the results of subjecting particles to forces. It includes celestial mechanics—the study of the motion of celestial objects.

Applied Mathematics

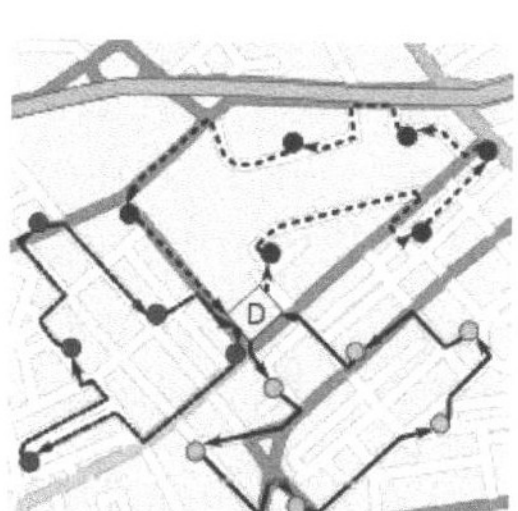

Efficient solutions to the vehicle routing problem require tools from combinatorial optimization and integer programming.

Applied mathematics is the application of mathematical methods by different fields such as science, engineering, business, computer science, and industry. Thus, applied mathematics is a combination of mathematical science and specialized knowledge. The term "applied mathematics" also describes the professional specialty in which mathematicians work on practical problems by formulating and studying mathematical models. In the past, practical applications have motivated the development of mathematical theories, which then became the subject of study in pure mathematics where abstract concepts are studied for their own sake. The activity of applied mathematics is thus intimately connected with research in pure mathematics.

History

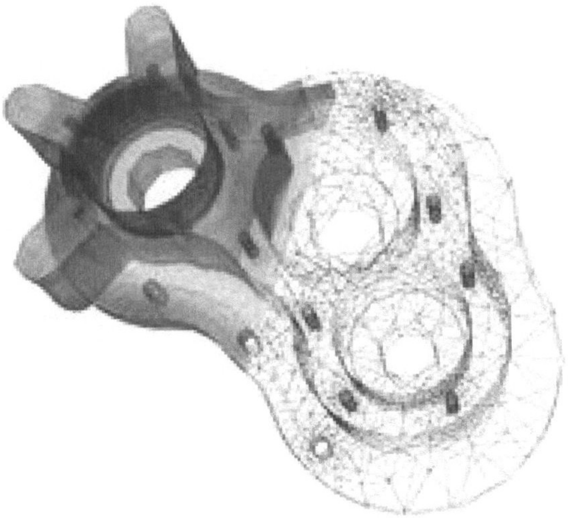

A numerical solution to the heat equation on a pump casing model using the finite element method.

Historically, applied mathematics consisted principally of applied analysis, most notably differential equations; approximation theory (broadly construed, to include representations, asymptotic methods, variational methods, and numerical analysis); and applied probability. These areas of mathematics related directly to the development of Newtonian physics, and in fact, the distinction between mathematicians and physicists was not sharply drawn before the mid-19th century. This history left a pedagogical legacy in the United States: until the early 20th century, subjects such as classical mechanics were often taught in applied mathematics departments at American universities rather than in physics departments, and fluid mechanics may still be taught in applied mathematics departments. Quantitative finance is now taught in mathematics departments across universities and mathematical finance is considered a full branch of applied mathematics. Engineeringand computer science departments have traditionally made use of applied mathematics.

Divisions

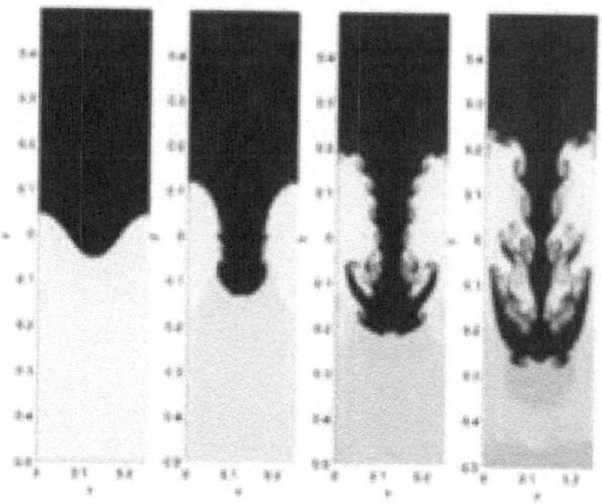

Fluid mechanics is often considered a branch of applied mathematics and mechanical engineering.

Today, the term "applied mathematics" is used in a broader sense. It includes the classical areas noted above as well as other areas that have become increasingly important in applications. Even fields such as number theory that are part of pure mathematics are now important in applications (such as cryptography), though they are not generally considered to be part of the field of applied mathematics per se. Sometimes, the term "applicable mathematics" is used to distinguish between the traditional applied mathematics that developed alongside physics and the many areas of mathematics that are applicable to real-world problems today.

There is no consensus as to what the various branches of applied mathematics are. Such categorizations are made difficult by the way mathematics and science change over time, and also by the way universities organize departments, courses, and degrees.

Many mathematicians distinguish between "applied mathematics," which is concerned with mathematical methods, and the "applications of mathematics" within science and engineering. A biologist using a population model and applying known mathematics would not be doingapplied mathematics, but rather using it; however, mathematical biologists have posed problems that have stimulated the growth of pure mathematics. Mathematicians such as Poincaré and Arnold deny the existence of "applied mathematics" and claim that there are only "applications of mathematics." Similarly, non-mathematicians blend applied mathematics and applications of mathematics. The use and development of mathematics to solve industrial problems is also called "industrial mathematics".

The success of modern numerical mathematical methods and software has led to the emergence of computational mathematics, computational science, and computational engineering, which use high-performance computing for the simulation of phenomena and the solution of problems in the sciences and engineering. These are often considered interdisciplinary.

Utility

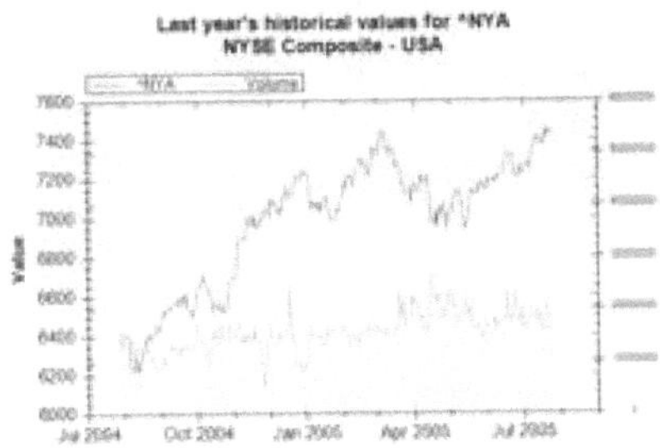

Mathematical finance is concerned with the modelling of financial markets.

Historically, mathematics was most important in the natural sciences and engineering. However, since World War II, fields outside the physical sciences have spawned the creation of new areas of mathematics, such as game theory and social choice theory, which grew out of economic considerations.

The advent of the computer has enabled new applications: studying and using the new computer technology itself (computer science) to study problems arising in other areas of science (computational science) as well as the mathematics of computation (for example, theoretical computer science, computer algebra, numerical analysis). Statistics is probably the most widespread mathematical science used in the social sciences, but other areas of mathematics, most notably economics, are proving increasingly useful in these disciplines.

Status in academic departments

Academic institutions are not consistent in the way they group and label courses, programs, and degrees in applied mathematics. At some schools, there is a single mathematics department, whereas others have separate departments for Applied Mathematics and (Pure) Mathematics. It is very common for Statistics departments to be separated at schools with graduate programs, but many undergraduate-only institutions include statistics under the mathematics department.

Many applied mathematics programs (as opposed to departments) consist of primarily cross-listed courses and jointly appointed faculty in departments representing applications. Some Ph.D. programs in applied mathematics require little or no coursework outside mathematics, while others require substantial coursework in a specific area of application. In some respects this difference reflects the distinction between "application of mathematics" and "applied mathematics".

Some universities in the UK host departments of Applied Mathematics and Theoretical Physics, but it is now much less common to have separate departments of pure and applied mathematics. A notable exception to this is the Department of Applied Mathematics and Theoretical Physics at the University of Cambridge, housing the Lucasian Professor of Mathematics whose past holders include Isaac Newton, Charles Babbage, James Lighthill, Paul Dirac and Stephen Hawking.

Schools with separate applied mathematics departments range from Brown University, which has a large Division of Applied Mathematics that offers degrees through the doctorate, to Santa Clara University, which offers only the M.S. in applied mathematics. Research universities dividing their mathematics department into pure and applied sections include MIT. Brigham Young University also has an Applied and Computational Emphasis (ACME), a program that allows student to graduate with a Mathematics degree, with an emphasis in Applied Math. Students in this program also learn another skill (Computer Science, Engineering, Physics, Pure Math, etc.) to supplement their applied math skills.

Associated mathematical sciences

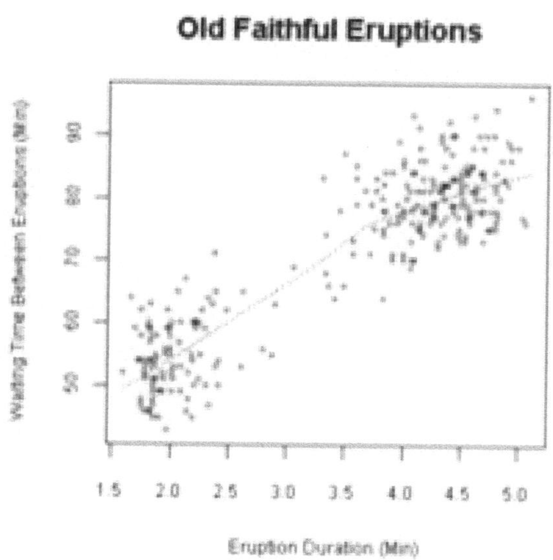

Applied mathematics has substantial overlap with statistics.

Applied mathematics is closely related to other mathematical sciences.

Scientific computing

Scientific computing includes applied mathematics (especially numerical analysis), computing science (especially high-performance computing), and mathematical modelling in a scientific discipline.

Computer science

Computer science relies on logic, algebra, graph theory, and combinatorics.

Operations research and management science

Operations research and management science are often taught in faculties of engineering, business, and public policy.

Statistics

Applied mathematics has substantial overlap with the discipline of statistics. Statistical theorists study and improve statistical procedures with mathematics, and statistical research often raises mathematical questions. Statistical theory relies on probability and decision theory, and makes extensive use of scientific computing, analysis, and optimization; for the design of experiments, statisticians use algebra and combinatorial design. Applied mathematicians and statisticians often work in a department of mathematical sciences (particularly at colleges and small universities).

Actuarial science

Actuarial science applies probability, statistics, and economic theory to assess risk in insurance, finance and other industries and professions.

Mathematical economics

Mathematical economics is the application of mathematical methods to represent theories and analyze problems in economics. The applied methods usually refer to nontrivial mathematical techniques or approaches. Mathematical economics is based on statistics, probability, mathematical programming (as well as other computational methods), operations research, game theory, and some methods from mathematical analysis. In this regard, it resembles (but is distinct from) financial mathematics, another part of applied mathematics.

According to the Mathematics Subject Classification (MSC), mathematical economics falls into the Applied mathematics/other classification of category 91:

Game theory, economics, social and behavioral sciences

with MSC2010 classifications for 'Game theory' at codes 91Axx and for 'Mathematical economics' at codes 91Bxx.

APPLICABLE MATHEMATICS

Applicable mathematics is a subdiscipline of applied mathematics, although there is no consensus as to a precise definition. Sometimes the term "applicable mathematics" is used to distinguish between the traditional applied mathematics that developed alongside physics and the many areas of mathematics that are applicable to real-world problems today.

Mathematicians often distinguish between "applied mathematics" on the one hand, and the "applications of mathematics" or "applicable mathematics" both within and outside of science and engineering, on the other. Some mathematicians emphasize the term applicable mathematics to separate or delineate the traditional applied areas from new applications arising from fields that were previously seen as pure mathematics. For example, from this viewpoint, an ecologist or geographer using population models and applying known mathematics would not be doing applied, but rather applicable, mathematics. Even fields such as number theory that are part of pure mathematics are now important in applications (such as cryptography), though they are not generally considered to be part of the field of applied mathematics per se.

Other authors prefer describing applicable mathematics as a union of "new" mathematical applications with the traditional fields of applied mathematics. With this outlook, the terms applied mathematics and applicable mathematics are thus interchangeable.

Other disciplines

The line between applied mathematics and specific areas of application is often blurred. Many universities teach mathematical and statistical courses outside the respective departments, in departments and areas including business, engineering, physics, chemistry, psychology, biology, computer science, scientific computation, and mathematical physics.

Discrete mathematics

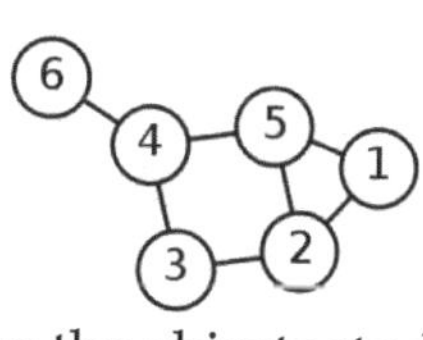

Graphs like this are among the objects studied by discrete mathematics, for their interesting mathematical properties, their usefulness as models of real-world problems, and their importance in developing computer algorithms.

Discrete mathematics is the study of mathematical structures that are fundamentally discrete rather than continuous. In contrast to real numbers that have the property of varying "smoothly", the objects studied in discrete mathematics – such as integers, graphs, and statements in logic – do not vary smoothly in this way, but have distinct, separated values. Discrete mathematics therefore excludes topics in "continuous mathematics" such as calculus or Euclidean geometry. Discrete objects can often be enumerated by integers. More formally, discrete mathematics has been characterized as the branch of mathematics dealing with countable sets (finite sets or sets with the same cardinality as the natural numbers). However, there is no exact definition of the term "discrete mathematics." Indeed, discrete mathematics is described less by what is included than by what is excluded: continuously varying quantities and related notions.

The set of objects studied in discrete mathematics can be finite or infinite. The term finite mathematics is sometimes applied to parts of the field of discrete mathematics that deals with finite sets, particularly those areas relevant to business.

Research in discrete mathematics increased in the latter half of the twentieth century partly due to the development of digital computerswhich operate in discrete steps and store data in discrete bits. Concepts and notations from discrete mathematics are useful in studying and describing objects and problems in branches of computer science, such as computer algorithms, programming languages, cryptography, automated theorem proving, and software development. Conversely, computer implementations are significant in applying ideas from discrete mathematics to real-world problems, such as in operations research.

Although the main objects of study in discrete mathematics are discrete objects, analytic methods from continuous mathematics are often employed as well.

In university curricula, "Discrete Mathematics" appeared in the 1980s, initially as a computer science support course; its contents were somewhat haphazard at the time. The curriculum has thereafter developed in conjunction with efforts by ACM and MAA into a course that is basically intended to develop mathematical maturity in freshmen; therefore it is nowadays a prerequisite for mathematics majors in some universities as well. Some high-school-level discrete mathematics textbooks have appeared as well. At this level, discrete mathematics is sometimes seen as a preparatory course, not unlike precalculus in this respect.

The Fulkerson Prize is awarded for outstanding papers in discrete mathematics.

Grand challenges, past and present

Much research in graph theory was motivated by attempts to prove that all maps, like this one, can be coloredusing only four colors so that no areas of the same color share an edge. Kenneth Appel and Wolfgang Haken proved this in 1976.

The history of discrete mathematics has involved a number of challenging problems which have focused attention within areas of the field. In graph theory, much research was motivated by attempts to prove the four color theorem, first stated in 1852, but not proved until 1976 (by Kenneth Appel and Wolfgang Haken, using substantial computer assistance).

In logic, the second problem on David Hilbert's list of open problems presented in 1900 was to prove that the axioms of arithmetic are consistent. Gödel's second incompleteness theorem, proved in 1931, showed that this was not possible – at least not within arithmetic itself. Hilbert's tenth problem was to determine whether a given polynomial Diophantine equation with integer coefficients has an integer solution. In 1970, Yuri Matiyasevich proved that this could not be done.

The need to break German codes in World War II led to advances in cryptography and theoretical computer science, with the first programmable digital electronic computer being developed at England's Bletchley Park with the guidance of Alan Turing and his seminal work, On Computable Numbers. At the same time, military requirements motivated advances in operations research. The Cold War meant that cryptography remained important, with fundamental advances such as public-key cryptography being developed in the following decades. Operations research remained important as a tool in business and project management, with the critical path method being developed in the 1950s. The telecommunicationindustry has also motivated advances in discrete mathematics, particularly in graph theory and information theory. Formal verification of statements in logic has been necessary for software development of safety-critical systems, and advances in automated theorem proving have been driven by this need.

Computational geometry has been an important part of the computer graphics incorporated into modern video games and computer-aided designtools.

Several fields of discrete mathematics, particularly theoretical computer science, graph theory, and combinatorics, are important in addressing the challenging bioinformatics problems associated with understanding the tree of life.

Currently, one of the most famous open problems in theoretical computer science is the P = NP problem, which involves the relationship between the complexity classes P and NP. The Clay Mathematics Institute has offered a $1 million USD prize for the first correct proof, along with prizes for six other mathematical problems.

Topics in discrete mathematics

Theoretical computer science

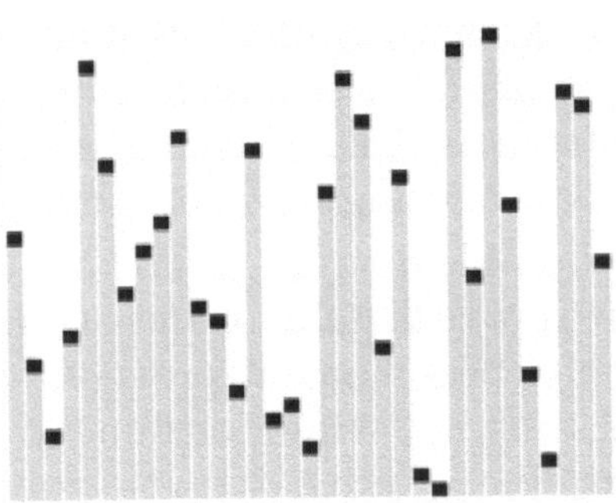

Complexity studies the time taken by algorithms, such as this sorting routine.

Theoretical computer science includes areas of discrete mathematics relevant to computing. It draws heavily on graph theory and mathematical logic. Included within theoretical computer science is the study of algorithms for computing mathematical results. Computabilitystudies what can be computed in principle, and has close ties to logic, while complexity studies the time, space, and other resources taken by computations. Automata theory and formal language theory are closely related to computability. Petri nets and process algebras are used to model computer systems, and methods from discrete mathematics are used in analyzing VLSI electronic circuits. Computational geometryapplies algorithms to geometrical problems, while computer image analysis applies them to representations of images. Theoretical computer science also includes the study of various continuous computational topics.

Information theory

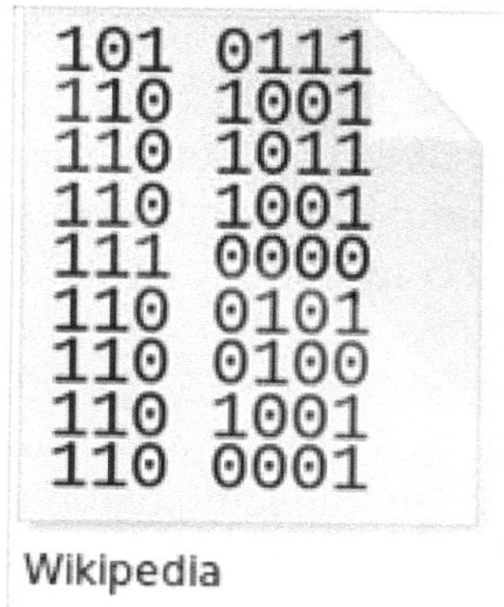

The ASCII codes for the word "Wikipedia", given here in binary, provide a way of representing the word in information theory, as well as for information-processing algorithms.

Information theory involves the quantification of information. Closely related is coding theory which is used to design efficient and reliable data transmission and storage methods. Information theory also includes continuous topics such as: analog signals, analog coding, analog encryption.

Logic

Logic is the study of the principles of valid reasoning and inference, as well as of consistency, soundness, and completeness. For example, in most systems of logic (but not in intuitionistic logic) Peirce's law $(((P\to Q)\to P)\to P)$ is a theorem. For classical logic, it can be easily verified with a truth table. The study of mathematical proof is particularly important in logic, and has applications to automated theorem proving and formal verification of software.

Logical formulas are discrete structures, as are proofs, which form finite trees or, more generally, directed acyclic graph structures (with each inference step combining one or more premise branches to give a single conclusion). The truth values of logical formulas usually form a finite set, generally restricted to two values: true and false, but logic can also be continuous-valued, e.g., fuzzy logic. Concepts such as infinite proof trees or infinite derivation trees have also been studied, e.g. infinitary logic.

Set theory

Set theory is the branch of mathematics that studies sets, which are collections of objects, such as {blue, white, red} or the (infinite) set of all prime numbers. Partially ordered sets and sets with other relations have applications in several areas.

In discrete mathematics, countable sets (including finite sets) are the main focus. The beginning of set theory as a branch of mathematics is usually marked by Georg Cantor's work distinguishing between different kinds of infinite set, motivated by the study of trigonometric series, and further development of the theory of infinite sets is outside the scope of discrete mathematics. Indeed, contemporary work in descriptive set theory makes extensive use of traditional continuous mathematics.

Combinatorics

Combinatorics studies the way in which discrete structures can be combined or arranged. Enumerative combinatorics concentrates on counting the number of certain combinatorial objects - e.g. the twelvefold way provides a unified framework for counting permutations, combinations and partitions. Analytic combinatorics concerns the enumeration (i.e., determining the number) of combinatorial structures using tools from complex analysis and probability theory. In contrast with enumerative combinatorics which uses explicit combinatorial formulae and generating functions to describe the results, analytic combinatorics aims at obtaining asymptotic formulae. Design theory is a study of combinatorial designs, which are collections of subsets with certain intersection properties. Partition theory studies various enumeration and asymptotic problems related to integer partitions, and is closely related to q-series, special functions and orthogonal polynomials. Originally a part of number theory and analysis, partition theory is now considered a part of combinatorics or an independent field. Order theory is the study of partially ordered sets, both finite and infinite.

Graph theory

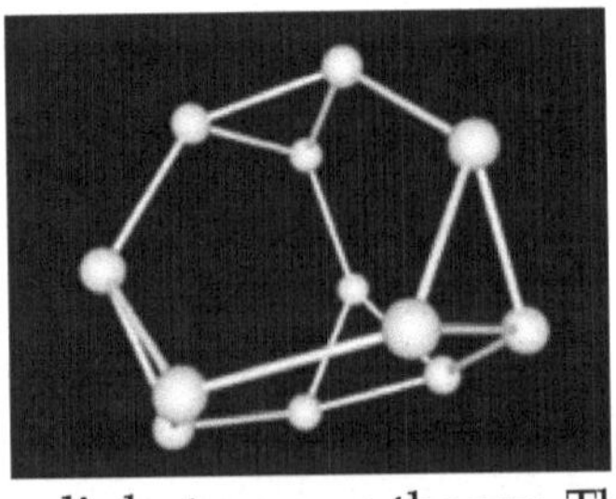

Graph theory has close links to group theory. This truncated tetrahedron graph is related to the alternating group A.

Graph theory, the study of graphs and networks, is often considered part of combinatorics, but has grown large enough and distinct enough, with its own kind of problems, to be regarded as a subject in its own right. Graphs are one of the prime objects of study in discrete mathematics. They are among the most ubiquitous models of both natural and human-made structures. They can model many types of relations and process dynamics in physical, biological and social systems. In computer science, they can represent networks of communication, data organization, computational devices, the flow of computation, etc. In mathematics, they are useful in geometry and certain parts of topology, e.g. knot theory. Algebraic graph theory has close links with group theory. There are also continuous graphs, however for the most part research in graph theory falls within the domain of discrete mathematics.

Probability

Discrete probability theory deals with events that occur in countable sample spaces. For example, count observations such as the numbers of birds in flocks comprise only natural number values {0, 1, 2, ...}. On the other hand, continuous observations such as the weights of birds comprise real number values and would typically be modeled by a continuous probability distribution such as the normal. Discrete probability distributions can be used to approximate continuous ones and vice versa. For highly constrained situations such as throwing dice or experiments with decks of cards, calculating the probability of events is basically enumerative combinatorics.

Number theory

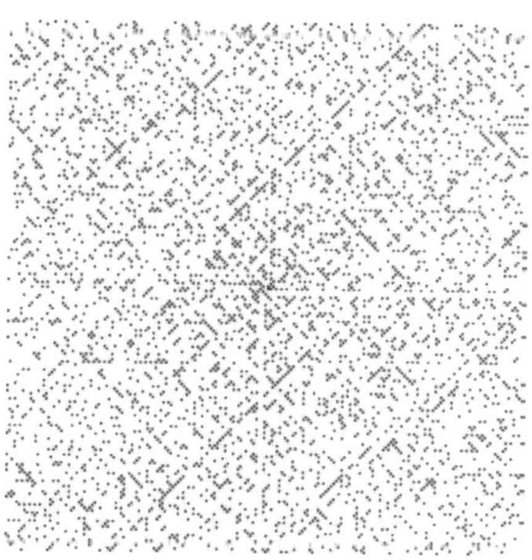

The Ulam spiral of numbers, with black pixels showing prime numbers. This diagram hints at patterns in the distribution of prime numbers.

Number theory is concerned with the properties of numbers in general, particularly integers. It has applications to cryptography and cryptanalysis, particularly with regard to modular arithmetic, diophantine equations, linear and quadratic congruences, prime numbers and primality testing. Other discrete aspects of number theory include geometry of numbers. In analytic number theory, techniques from continuous mathematics are also used. Topics that go beyond discrete objects include transcendental numbers, diophantine approximation, p-adic analysisand function fields.

Algebra

Algebraic structures occur as both discrete examples and continuous examples. Discrete algebras include: boolean algebra used in logic gatesand programming; relational algebra used in databases; discrete and finite versions of groups, rings and fields are important in algebraic coding theory; discrete semigroups and monoids appear in the theory of formal languages.

Calculus of finite differences, discrete calculus or discrete analysis

A function defined on an interval of the integers is usually called a sequence. A sequence could be a finite sequence from a data source or an infinite sequence from a discrete dynamical system. Such a discrete function could be defined explicitly by a list (if its domain is finite), or by a formula for its general term, or it could be given implicitly by a recurrence relation or difference equation. Difference equations are similar to a differential equations, but replace differentiation by taking the difference between adjacent terms; they can be used to approximate differential equations or (more often) studied in their own right. Many questions and methods concerning differential equations have counterparts for difference equations. For instance, where there are integral transforms in harmonic analysis for studying continuous functions or analogue signals, there are discrete transforms for discrete functions or digital signals. As well as the discrete metric there are more general discrete or finite metric spaces and finite topological spaces.

Geometry

Computational geometry applies computer algorithms to representations of geometrical objects.

Discrete geometry and combinatorial geometry are about combinatorial properties of discrete collections of geometrical objects. A long-standing topic in discrete geometry is tiling of the plane. Computational geometry applies algorithms to geometrical problems.

Topology

Although topology is the field of mathematics that formalizes and generalizes the intuitive notion of "continuous deformation" of objects, it gives rise to many discrete topics; this can be attributed in part to the focus on topological invariants, which themselves usually take discrete values. See combinatorial topology, topological graph theory, topological combinatorics, computational topology, discrete topological space, finite topological space, topology (chemistry).

OPERATIONS RESEARCH

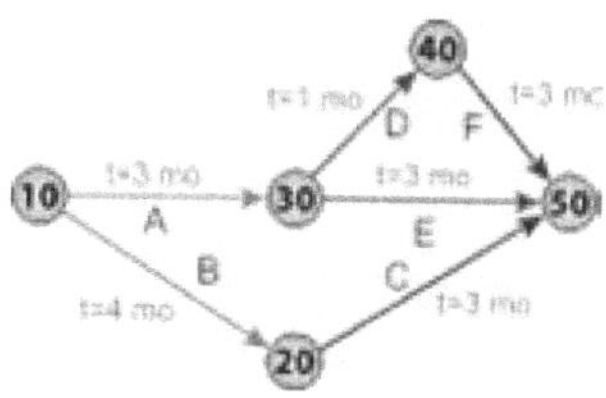

PERT charts like this provide a project management technique based on graph theory.

Operations research provides techniques for solving practical problems in engineering, business, and other fields — problems such as allocating resources to maximize profit, or scheduling project activities to minimize risk. Operations research techniques include linear programming and other areas of optimization, queuing theory, scheduling theory, network theory. Operations research also includes continuous topics such as continuous-time Markov process, continuous-time martingales, process optimization, and continuous and hybrid control theory.

Game theory, decision theory, utility theory, social choice theory[edit

	Cooperate	Defect
Cooperate	−1, −1	−10, 0
Defect	0, −10	−5, −5

Payoff matrix for the Prisoner's dilemma, a common example in game theory. One player chooses a row, the other a column; the resulting pair gives their payoffs

Decision theory is concerned with identifying the values, uncertainties and other issues relevant in a given decision, its rationality, and the resulting optimal decision. Utility theory is about measures of the relative economic satisfaction from, or desirability of, consumption of various goods and services.

Social choice theory is about voting. A more puzzle-based approach to voting is ballot theory.

Game theory deals with situations where success depends on the choices of others, which makes choosing the best course of action more complex. There are even continuous games, see differential game. Topics include auction theory and fair division.

Discretization

Discretization concerns the process of transferring continuous models and equations into discrete counterparts, often for the purposes of making calculations easier by using approximations. Numerical analysis provides an important example.

Discrete analogues of continuous mathematics

There are many concepts in continuous mathematics which have discrete versions, such as discrete calculus, discrete probability distributions, discrete Fourier transforms, discrete geometry, discrete logarithms, discrete differential geometry, discrete exterior calculus, discrete Morse theory, difference equations, discrete dynamical systems, and discrete vector measures.

In applied mathematics, discrete modelling is the discrete analogue of continuous modelling. In discrete modelling, discrete formulae are fit to data. A common method in this form of modelling is to use recurrence relation.

COMPUTATIONAL MATHEMATICS

A black and white rendition of the Yale Babylonian Collection's Tablet YBC 7289 (c. 1800–1600 BCE), showing a Babylonian approximation to the square root of 2 (1 24 51 10 w: sexagesimal) in the context of Pythagoras'

Theorem for an isosceles triangle. The tablet also gives an example where one side of the square is 30, and the resulting diagonal is 42 25 35 or 42.4263888.

Computational mathematics may refer to two different aspect of the relation between computing and mathematics.

Computational applied mathematics consists roughly of using mathematics for allowing and improving computer computation in applied mathematics. Computational mathematics may also refer to the use of computers for mathematics itself. This includes the use of computers for mathematical computations (computer algebra), the study of what can (and cannot) be computerized in mathematics (effective methods), which computations may be done with present technology (complexity theory), and which proofs can be done on computers (proof assistants).

Both aspects of computational mathematics involves mathematical research in mathematics as well as in areas of science where computingplays a central and essential role—that, is almost all sciences—, and emphasize algorithms, numerical methods, and symbolic computations.

Areas of computational mathematics

Computational mathematics emerged as a distinct part of applied mathematics by the early 1950s. Currently, computational mathematics can refer to or include:

- computational science, also known as scientific computation or computational engineering
- solving mathematical problems by computer simulation as opposed to analytic methods of applied mathematics
- numerical methods used in scientific computation, for example numerical linear algebra and numerical solution of partial differential equations
- stochastic methods, such as Monte Carlo methods and other representations of uncertainty in scientific computation
- the mathematics of scientific computation, , in particular numerical analysis, the theory of numerical methods
- computational complexity
- computer algebra and computer algebra systems
- computer-assisted research in various areas of mathematics, such as logic (automated theorem proving), discrete mathematics, combinatorics), number theory, and computational algebraic topology
- cryptography and computer security, which involve, in particular, research on primality testing, factorization, elliptic curves, and mathematics of blockchain.

- computational linguistics, the use of mathematical and computer techniques in natural languages
- computational algebraic geometry
- computational group theory
- computational geometry
- computational number theory
- computational topology
- computational statistics
- algorithmic information theory
- algorithmic game theory
- mathematical economics, the use of mathematics in economics, finance and, to certain extents, of accounting.

3

Abstract Algebra

In algebra, which is a broad division of mathematics, abstract algebra (occasionally called modern algebra) is the study of algebraic structures. Algebraic structures include groups, rings, fields, modules, vector spaces, lattices, and algebras. The term abstract algebra was coined in the early 20th century to distinguish this area of study from the other parts of algebra.

Algebraic structures, with their associated homomorphisms, form mathematical categories. Category theory is a formalism that allows a unified way for expressing properties and constructions that are similar for various structures.

Universal algebra is a related subject that studies types of algebraic structures as single objects. For example, the structure of groups is a single object in universal algebra, which is called variety of groups.

History of Abstract algebra

As in other parts of mathematics, concrete problems and examples have played important roles in the development of abstract algebra. Through the end of the nineteenth century, many – perhaps most – of these problems were in some way related to the theory of algebraic equations. Major themes include:

- Solving of systems of linear equations, which led to linear algebra
- Attempts to find formulas for solutions of general polynomial equations of higher degree that resulted in discovery of groups as abstract manifestations of symmetry
- Arithmetical investigations of quadratic and higher degree forms and diophantine equations, that directly produced the notions of a ring and ideal.

Numerous textbooks in abstract algebra start with axiomatic definitions of various algebraic structures and then proceed to establish their properties.

This creates a false impression that in algebra axioms had come first and then served as a motivation and as a basis of further study. The true order of historical development was almost exactly the opposite. For example, the hypercomplex numbers of the nineteenth century had kinematic and physical motivations but challenged comprehension. Most theories that are now recognized as parts of algebra started as collections of disparate facts from various branches of mathematics, acquired a common theme that served as a core around which various results were grouped, and finally became unified on a basis of a common set of concepts. An archetypical example of this progressive synthesis can be seen in the history of group theory.

Early group theory

There were several threads in the early development of group theory, in modern language loosely corresponding to number theory, theory of equations, and geometry.

Leonhard Euler considered algebraic operations on numbers modulo an integer, modular arithmetic, in his generalization of Fermat's little theorem. These investigations were taken much further by Carl Friedrich Gauss, who considered the structure of multiplicative groups of residues mod n and established many properties of cyclic and more general abelian groups that arise in this way. In his investigations of composition of binary quadratic forms, Gauss explicitly stated the associative law for the composition of forms, but like Euler before him, he seems to have been more interested in concrete results than in general theory. In 1870, Leopold Kronecker gave a definition of an abelian group in the context of ideal class groups of a number field, generalizing Gauss's work; but it appears he did not tie his definition with previous work on groups, particularly permutation groups. In 1882, considering the same question, Heinrich M. Weber realized the connection and gave a similar definition that involved the cancellation property but omitted the existence of the inverse element, which was sufficient in his context (finite groups).

Permutations were studied by Joseph-Louis Lagrange in his 1770 paper Réflexions sur la résolution algébrique des équations (Thoughts on the algebraic solution of equations) devoted to solutions of algebraic equations, in which he introduced Lagrange resolvents. Lagrange's goal was to understand why equations of third and fourth degree admit formulas for solutions, and he identified as key objects permutations of the roots. An important novel step taken by Lagrange in this paper was the abstract view of the roots, i.e. as symbols and not as numbers. However, he did not consider composition of permutations. Serendipitously, the first edition of

Edward Waring's Meditationes Algebraicae (Meditations on Algebra) appeared in the same year, with an expanded version published in 1782. Waring proved the fundamental theorem of symmetric polynomials, and specially considered the relation between the roots of a quartic equation and its resolvent cubic. Mémoire sur la résolution des équations (Memoire on the Solving of Equations) of Alexandre Vandermonde (1771) developed the theory of symmetric functions from a slightly different angle, but like Lagrange, with the goal of understanding solvability of algebraic equations.

Kronecker claimed in 1888 that the study of modern algebra began with this first paper of Vandermonde. Cauchy states quite clearly that Vandermonde had priority over Lagrange for this remarkable idea, which eventually led to the study of group theory.

Paolo Ruffini was the first person to develop the theory of permutation groups, and like his predecessors, also in the context of solving algebraic equations. His goal was to establish the impossibility of an algebraic solution to a general algebraic equation of degree greater than four. En route to this goal he introduced the notion of the order of an element of a group, conjugacy, the cycle decomposition of elements of permutation groups and the notions of primitive and imprimitive and proved some important theorems relating these concepts, such as

if G is a subgroup of S_5 whose order is divisible by 5 then G contains an element of order 5.

Note, however, that he got by without formalizing the concept of a group, or even of a permutation group. The next step was taken by Évariste Galois in 1832, although his work remained unpublished until 1846, when he considered for the first time what is now called the closure property of a group of permutations, which he expressed as if in such a group one has the substitutions S and T then one has the substitution ST.

The theory of permutation groups received further far-reaching development in the hands of Augustin Cauchy and Camille Jordan, both through introduction of new concepts and, primarily, a great wealth of results about special classes of permutation groups and even some general theorems. Among other things, Jordan defined a notion of isomorphism, still in the context of permutation groups and, incidentally, it was he who put the term group in wide use.

The abstract notion of a group appeared for the first time in Arthur Cayley's papers in 1854. Cayley realized that a group need not be a permutation group (or even finite), and may instead consist of matrices, whose algebraic properties, such as multiplication and inverses, he systematically investigated in succeeding years. Much later Cayley would revisit the question whether abstract groups were more general than permutation groups, and establish that, in fact, any group is isomorphic to a group of permutations.

Modern algebra

The end of the 19th and the beginning of the 20th century saw a tremendous shift in the methodology of mathematics. Abstract algebra emerged around the start of the 20th century, under the name modern algebra. Its study was part of the drive for more intellectual rigor in mathematics. Initially, the assumptions in classical algebra, on which the whole of mathematics (and major parts of the natural sciences) depend, took the form of axiomatic systems. No longer satisfied with establishing properties of concrete objects, mathematicians started to turn their attention to general theory. Formal definitions of certain algebraic structures began to emerge in the 19th century. For example, results about various groups of permutations came to be seen as instances of general theorems that concern a general notion of an abstract group. Questions of structure and classification of various mathematical objects came to forefront.

These processes were occurring throughout all of mathematics, but became especially pronounced in algebra. Formal definition through primitive operations and axioms were proposed for many basic algebraic structures, such as groups, rings, and fields. Hence such things as group theory and ring theory took their places in pure mathematics. The algebraic investigations of general fields by Ernst Steinitz and of commutative and then general rings by David Hilbert, Emil Artin and Emmy Noether, building up on the work of Ernst Kummer, Leopold Kronecker and Richard Dedekind, who had considered ideals in commutative rings, and of Georg Frobenius and Issai Schur, concerning representation theory of groups, came to define abstract algebra. These developments of the last quarter of the 19th century and the first quarter of 20th century were systematically exposed in Bartel van der Waerden's Moderne algebra, the two-volume monograph published in 1930–1931 that forever changed for the mathematical world the meaning of the word algebra from the theory of equations to the theory of algebraic structures.

Basic concepts

By abstracting away various amounts of detail, mathematicians have defined various algebraic structures that are used in many areas of mathematics. For instance, almost all systems studied are sets, to which the theorems of set theory apply. Those sets that have a certain binary operation defined on them form magmas, to which the concepts concerning magmas, as well those concerning sets, apply. We can add additional constraints on the algebraic structure, such as associativity (to form semigroups); identity, and inverses (to form groups); and other more complex structures. With additional structure, more theorems could be proved, but the generality is reduced. The "hierarchy" of algebraic objects (in terms of generality) creates a hierarchy of the corresponding theories: for instance, the theorems of group theory may be used when studying rings (algebraic objects that have two binary operations with certain axioms) since a ring is a group over one of its operations. In general there is a balance between the amount of generality and the richness of the theory: more general structures have usually fewer nontrivial theorems and fewer applications.

Examples of algebraic structures with a single binary operation are:

- Magma
- Quasigroup
- Monoid
- Semigroup
- Group

Examples involving several operations include:

- Ring
- Field
- Module
- Vector space
- Algebra over a field
- Associative algebra
- Lie algebra
- Lattice
- Boolean algebra

Algebraic structure

In mathematics, and more specifically in abstract algebra, an algebraic structure on a set A (called carrier set or underlying set) is a collection of finitary operations on A; the set A with this structure is also called an algebra.

Examples of algebraic structures include groups, rings, fields, and lattices. More complex structures can be defined by introducing multiple operations, different underlying sets, or by altering the defining axioms. Examples of more complex algebraic structures include vector spaces, modules, and algebras.

The properties of specific algebraic structures are studied in abstract algebra. The general theory of algebraic structures has been formalized in universal algebra. The language of category theory is used to express and study relationships between different classes of algebraic and non-algebraic objects. This is because it is sometimes possible to find strong connections between some classes of objects, sometimes of different kinds. For example, Galois theory establishes a connection between certain fields and groups: two algebraic structures of different kinds.

INTRODUCTION

Addition and multiplication on numbers are the prototypical example of an operation that combines two elements of a set to produce a third. These operations obey several algebraic laws. For example, $a + (b + c) = (a + b) + c$ and $a(bc) = (ab)c$, both examples of the associative law. Also $a + b = b + a$, and $ab = ba$, the commutative law. Many systems studied by mathematicians have operations that obey some, but not necessarily all, of the laws of ordinary arithmetic. For example, rotations of objects in three-dimensional space can be combined by performing the first rotation and then applying the second rotation to the object in its new orientation. This operation on rotations obeys the associative law, but can fail the commutative law.

Mathematicians give names to sets with one or more operations that obey a particular collection of laws, and study them in the abstract as algebraic structures. When a new problem can be shown to follow the laws of one of these algebraic structures, all the work that has been done on that category in the past can be applied to the new problem.

In full generality, algebraic structures may involve an arbitrary number of sets and operations that can combine more than two elements (higher arity), but this article focuses on binary operations on one or two sets. The examples here are by no means a complete list, but they are meant to be a representative list and include the most common structures. Longer lists of algebraic structures may be found in the external links and within Category:Algebraic structures. Structures are listed in approximate order of increasing complexity.

Examples

One set with operations

Simple structures: no binary operation:

- Set: a degenerate algebraic structure S having no operations.
- Pointed set: S has one or more distinguished elements, often 0, 1, or both.
- Unary system: S and a single unary operation over S.
- Pointed unary system: a unary system with S a pointed set.

Group-like structures: one binary operation. The binary operation can be indicated by any symbol, or with no symbol (juxtaposition) as is done for ordinary multiplication of real numbers.

- Magma or groupoid: S and a single binary operation over S.
- Semigroup: an associative magma.
- Monoid: a semigroup with Identity element.
- Group: a monoid with a unary operation (inverse), giving rise to inverse elements.
- Abelian group: a group whose binary operation is commutative.
- Semilattice: a semigroup whose operation is idempotent and commutative. The binary operation can be called either meet or join.
- Quasigroup: a magma obeying the latin square property. A quasigroup may also be represented using three binary operations.
- Loop: a quasigroup with identity.

Ring-like structures or Ringoids: two binary operations, often called addition and multiplication, with multiplication distributing over addition.

- Semiring: a ringoid such that S is a monoid under each operation. Addition is typically assumed to be commutative and associative, and the monoid product is assumed to distribute over the addition on both sides, and the additive identity satisfies 0/ x = 0 for all x.
- Near-ring: a semiring whose additive monoid is a (not necessarily abelian) group.
- Ring: a semiring whose additive monoid is an abelian group.
- Lie ring: a ringoid whose additive monoid is an abelian group, but whose multiplicative operation satisfies the Jacobi identity rather than associativity.
- Boolean ring: a commutative ring with idempotent multiplication operation.
- Field: a commutative ring which contains a multiplicative inverse for every nonzero element
- Kleene algebras: a semiring with idempotent addition and a unary operation, the Kleene star, satisfying additional properties.
- *-algebra: a ring with an additional unary operation (*) satisfying additional properties.

Lattice structures: two or more binary operations, including operations called meet and join, connected by the absorption law.

- Complete lattice: a lattice in which arbitrary meet and joins exist.
- Bounded lattice: a lattice with a greatest element and least element.
- Complemented lattice: a bounded lattice with a unary operation, complementation, denoted by postfix ¥". The join of an element with its complement is the greatest element, and the meet of the two elements is the least element.

- Modular lattice: a lattice whose elements satisfy the additional modular identity.
- Distributive lattice: a lattice in which each of meet and join distributes over the other. Distributive lattices are modular, but the converse does not hold.
- Boolean algebra: a complemented distributive lattice. Either of meet or join can be defined in terms of the other and complementation. This can be shown to be equivalent with the ring-like structure of the same name above.
- Heyting algebra: a bounded distributive lattice with an added binary operation, relative pseudo-complement, denoted by infix →, and governed by the axiomsx/ →/ x = 1,x/ (x/ →/ y) = x/ y ,y/ (x/ →/ y) = y ,x/ →/ (y/ z) = (x/ →/ y)/ (x/ →/ z).

Arithmetics: two binary operations, addition and multiplication. S is an infinite set. Arithmetics are pointed unary systems, whose unary operation is injective successor, and with distinguished element 0.

- Robinson arithmetic. Addition and multiplication are recursively defined by means of successor. 0 is the identity element for addition, and annihilates multiplication. Robinson arithmetic is listed here even though it is a variety, because of its closeness to Peano arithmetic.
- Peano arithmetic. Robinson arithmetic with an axiom schema of induction. Most ring and field axioms bearing on the properties of addition and multiplication are theorems of Peano arithmetic or of proper extensions thereof.

Two sets with operations

Module-like structures: composite systems involving two sets and employing at least two binary operations.

- Group with operators: a group G with a set Ω and a binary operation Ω/ ×/ G → G satisfying certain axioms.
- Module: an abelian group M and a ring R acting as operators on M. The members of R are sometimes called scalars, and the binary operation of scalar multiplication is a function R/ ×/ M→M, which satisfies several axioms. Counting the ring operations these systems have at least three operations.
- Vector space: a module where the ring R is a division ring or field.
- Graded vector space: a vector space with a direct sum decomposition breaking the space into "grades".
- Quadratic space: a vector space V over a field F with a function from V into F satisfying certain properties. Every quadratic space is also an inner product space (see below).

Algebra-like structures: composite system defined over two sets, a ring R and an R-module M equipped with an operation called multiplication. This can be viewed as a system with five binary operations: two operations on R, two on M and one involving both R and M.

- Algebra over a ring (also R-algebra): a module over a commutative ring R, which also carries a multiplication operation that is compatible with the module structure. This includes distributivity over addition and linearity with respect to multiplication by elements of R. The theory of an algebra over a field is especially well developed.
- Associative algebra: an algebra over a ring such that the multiplication is associative.
- Nonassociative algebra: a module over a commutative ring, equipped with a ring multiplication operation that is not necessarily associative. Often associativity is replaced with a different identity, such as alternation, the Jacobi identity, or the Jordan identity.
- Coalgebra: a vector space with a "comultiplication" defined dually to that of associative algebras.
- Lie algebra: a special type of nonassociative algebra whose product satisfies the Jacobi identity.
- Lie coalgebra: a vector space with a "comultiplication" defined dually to that of Lie algebras.
- Graded algebra: a graded vector space with an algebra structure compatible with the grading. The idea is that if the grades of two elements a and b are known, then the grade of ab is known, and so the location of the product ab is determined in the decomposition.
- Inner product space: an F vector space V with a sesquilinear binary operation $V \times V \rightarrow F$.

  Four or more binary operations:
- Bialgebra: an associative algebra with a compatible coalgebra structure.
- Lie bialgebra: a Lie algebra with a compatible bialgebra structure.
- Hopf algebra: a bialgebra with a connection axiom (antipode).
- Clifford algebra: a graded associative algebra equipped with an exterior product from which may be derived several possible inner products. Exterior algebras and geometric algebras are special cases of this construction.

Hybrid structures

Algebraic structures can also coexist with added structure of non-algebraic nature, such as partial order or a topology. The added structure must be compatible, in some sense, with the algebraic structure.

- Topological group: a group with a topology compatible with the group operation.

- Lie group: a topological group with a compatible smooth manifold structure.
- Ordered groups, ordered rings and ordered fields: each type of structure with a compatible partial order.
- Archimedean group: a linearly ordered group for which the Archimedean property holds.
- Topological vector space: a vector space whose M has a compatible topology.
- Normed vector space: a vector space with a compatible norm. If such a space is complete (as a metric space) then it is called a Banach space.
- Hilbert space: an inner product space over the real or complex numbers whose inner product gives rise to a Banach space structure.
- Vertex operator algebra
- Von Neumann algebra: a *-algebra of operators on a Hilbert space equipped with the weak operator topology.

Binary operation

A binary operation ‘‴ is a calculation that combines the arguments x and y to x ‘‴ y.

In mathematics, a binary operation or dyadic operation is a calculation that combines two elements (called operands) to produce another element. More formally, a binary operation is an operation of arity two.

More specifically, a binary operation on a set is a binary operation whose two domains and the codomain are the same set. Examples include the familiar arithmetic operations of addition, subtraction, multiplication. Other examples are readily found in different areas of mathematics, such as vector addition, matrix multiplication and conjugation in groups.

However, a binary operation may also involve several sets. For example, scalar multiplication of vector spaces takes a scalar and a vector to produce a vector, and scalar product takes two vectors to produce a scalar.

Binary operations are the keystone of most algebraic structures, that are studied in algebra, and used in all mathematics, such as fields, groups, monoids, rings, algebras, and many more.

Terminology

More precisely, a binary operation on a set S is a map which sends elements of the Cartesian product S × S to S:

$f : S \times S \to S$.

Because the result of performing the operation on a pair of elements of S is again an element of S, the operation is called a closed binary operation on S (or sometimes expressed as having the property of closure). If f is not a function, but is instead a partial function, it is called a partial binary operation. For instance, division of real numbers is a partial binary operation, because one can't divide by zero: a/0 is not defined for any real a. However, both in universal algebra and model theory the binary operations considered are defined on all of S × S.

Sometimes, especially in computer science, the term is used for any binary function.

Properties and examples

Typical examples of binary operations are the addition (+) and multiplication (×) of numbers and matrices as well as composition of functions on a single set. For instance,

- On the set of real numbers R, f(a, b) = a + b is a binary operation since the sum of two real numbers is a real number.
- On the set of natural numbers N, f(a, b) = a + b is a binary operation since the sum of two natural numbers is a natural number. This is a different binary operation than the previous one since the sets are different.
- On the set M(2,R) of 2 × 2 matrices with real entries, f(A, B) = A + B is a binary operation since the sum of two such matrices is a 2 × 2 matrix.
- On the set M(2,R) of 2 × 2 matrices with real entries, f(A, B) = AB is a binary operation since the product of two such matrices is a 2 × 2 matrix.
- For a given set C, let S be the set of all functions h : C→C. Define f : S × S→S by f(h_1, h_2)(c) = (h_1 '" h_2) (c) = h_1(h_2(c)) for all c " C, the composition of the two functions h_1 and h_2 in S. Then f is a binary operation since the composition of the two functions is again a function on the set C (that is, a member of S).

Many binary operations of interest in both algebra and formal logic are commutative, satisfying f(a, b) = f(b, a) for all elements a and b in S, or associative, satisfying f(f(a, b), c) = f(a, f(b, c)) for all a, b and c in S. Many also have identity elements and inverse elements.

The first three examples above are commutative and all of the above examples are associative.

On the set of real numbers R, subtraction, that is, f(a, b) = a " b, is a binary operation which is not commutative since, in general, a " b `" b " a. It is also not associative, since, in general, a " (b " c) `" (a " b) " c; for instance, 1 " (2 " 3) = 2 but (1 " 2) " 3 = "4.

On the set of natural numbers N, the binary operation exponentiation, f(a,b) = a^b, is not commutative since, a^b '" b^a (cf. Equation x¸ = yã), and is also not associative since f(f(a, b), c) '" f(a, f(b, c)). For instance, with a = 2, b = 3 and c = 2, f(2^3,2) = f(8,2) = 8^2 = 64, but f(2,3^2) = f(2,9) = 2^9 = 512. By changing the set N to the set of integers Z, this binary operation becomes a partial binary operation since it is now undefined when a = 0 and b is any negative integer. For either set, this operation has a right identity (which is 1) since f(a, 1) = a for all a in the set, which is not an identity (two sided identity) since f(1, b) '" b in general.

Division (/), a partial binary operation on the set of real or rational numbers, is not commutative or associative. Tetration ('!'!), as a binary operation on the natural numbers, is not commutative or associative and has no identity element.

NOTATION

Binary operations are often written using infix notation such as a " b, a + b, a · b or (by juxtaposition with no symbol) ab rather than by functional notation of the form f(a, b). Powers are usually also written without operator, but with the second argument as superscript.

Binary operations sometimes use prefix or (probably more often) postfix notation, both of which dispense with parentheses. They are also called, respectively, Polish notation and reverse Polish notation.

Pair and tuple

A binary operation, ab, depends on the ordered pair (a, b) and so (ab)c (where the parentheses here mean first operate on the ordered pair (a, b) and then operate on the result of that using the ordered pair ((ab), c)) depends in general on the ordered pair ((a, b), c). Thus, for the general, non-associative case, binary operations can be represented with binary trees.

However:

- If the operation is associative, (ab)c = a(bc), then the value of (ab)c depends only on the tuple (a, b, c).
- If the operation is commutative, ab = ba, then the value of (ab)c depends only on { {a, b}, c}, where braces indicate multisets.
- If the operation is both associative and commutative then the value of (ab)c depends only on the multiset {a, b, c}.
- If the operation is associative, commutative and idempotent, aa = a, then the value of (ab)c depends only on the set {a, b, c}.

Binary operations as ternary relations

A binary operation f on a set S may be viewed as a ternary relation on S, that is, the set of triples (a, b, f(a,b)) in S × S × S for all a and b in S.

External binary operations

An external binary operation is a binary function from K × S to S. This differs from a binary operation on a set in the sense in that K need not be S; its elements come from outside.

An example of an external binary operation is scalar multiplication in linear algebra. Here K is a field and S is a vector space over that field.

An external binary operation may alternatively be viewed as an action; K is acting on S.

The dot product of two vectors maps from S × S to K, where K is a field and S is a vector space over K. It depends on authors whether it is considered as a binary operation.

UNIVERSAL ALGEBRA

Algebraic structures are defined through different configurations of axioms. Universal algebra abstractly studies such objects. One major dichotomy is between structures that are axiomatized entirely by identities and structures that are not. If all axioms defining a class of algebras are identities, then this class is a variety (not to be confused with algebraic varieties of algebraic geometry).

Identities are equations formulated using only the operations the structure allows, and variables that are tacitly universally quantified over the relevant universe. Identities contain no connectives, existentially quantified variables, or relations of any kind other than the allowed operations. The study of varieties is an important part of universal algebra. An algebraic structure in a variety may be understood as the quotient algebra of term algebra (also called "absolutely free algebra") divided by the equivalence relations generated by a set of identities. So, a collection of functions with given signatures generate a free algebra, the term algebra T. Given a set of equational identities (the axioms), one may consider their symmetric, transitive closure E. The quotient algebra T/E is then the algebraic structure or variety. Thus, for example, groups have a signature containing two operators: the multiplication operator m, taking two arguments, and the inverse operator i, taking one argument, and the identity element e, a constant, which may be considered an operator that takes zero arguments. Given a (countable) set of variables x, y, z, etc. the term algebra is the collection of all possible terms involving m, i, e and the variables; so for example, m(i(x), m(x,m(y,e))) would be an element of the term algebra. One of the axioms defining a group is the identity m(x, i(x)) = e; another is m(x,e) = x. The axioms can be represented as trees. These equations induce equivalence classes on the free algebra; the quotient algebra then has the algebraic structure of a group.

Some structures do not form varieties, because either:

1. It is necessary that 0 ‘‴ 1, 0 being the additive identity element and 1 being a multiplicative identity element, but this is a nonidentity;

2. Structures such as fields have some axioms that hold only for nonzero members of S. For an algebraic structure to be a variety, its operations must be defined for all members of S; there can be no partial operations.

Structures whose axioms unavoidably include nonidentities are among the most important ones in mathematics, e.g., fields and division rings. Structures with nonidentities present challenges varieties do not. For example, the direct product of two fields is not a field, because (1 , 0) Å" (0 , 1) = (0 , 0) {\displaystyle (1,0)\cdot (0,1)=(0,0)} $(1,0)\cdot(0,1)=(0,0)$, but fields do not have zero divisors.

Category theory

Category theory is another tool for studying algebraic structures (see, for example, Mac Lane 1998). A category is a collection of objects with associated morphisms. Every algebraic structure has its own notion of homomorphism, namely any function compatible with the operation(s) defining the structure. In this way, every algebraic structure gives rise to a category. For example, the category of groups has all groups as objects and all group homomorphisms as morphisms. This concrete category may be seen as a category of sets with added category-theoretic structure. Likewise, the category of topological groups (whose morphisms are the continuous group homomorphisms) is a category of topological spaces with extra structure. A forgetful functor between categories of algebraic structures "forgets" a part of a structure.

There are various concepts in category theory that try to capture the algebraic character of a context, for instance

- algebraic category
- essentially algebraic category
- presentable category
- locally presentable category
- monadic functors and categories
- universal property.

DIFFERENT MEANINGS OF "STRUCTURE"

In a slight abuse of notation, the word "structure" can also refer to just the operations on a structure, instead of the underlying set itself. For example, the sentence, "We have defined a ring structure on the set A {\displaystyle A} A," means that we have defined ring operations on the set A {\displaystyle A} A. For another example, the group (Z , +) {\displaystyle (\mathbb {Z} ,+)} $(\mathbb{Z},+)$ can be seen as a set Z {\displaystyle \mathbb {Z} } $\mathbb{Z}$ that is equipped with an algebraic structure, namely the operation + {\displaystyle +} $+$.

Applications

Because of its generality, abstract algebra is used in many fields of mathematics and science. For instance, algebraic topology uses algebraic objects to study topologies. The Poincaré conjecture, proved in 2003, asserts that the fundamental group of a manifold, which encodes information about connectedness, can be used to determine whether a manifold is a sphere or not. Algebraic number theory studies various number rings that generalize the set of integers. Using tools of algebraic number theory, Andrew Wiles proved Fermat's Last Theorem.

In physics, groups are used to represent symmetry operations, and the usage of group theory could simplify differential equations. In gauge theory, the requirement of local symmetry can be used to deduce the equations describing a system. The groups that describe those symmetries are Lie groups, and the study of Lie groups and Lie algebras reveals much about the physical system; for instance, the number of force carriers in a theory is equal to the dimension of the Lie algebra, and these bosons interact with the force they mediate if the Lie algebra is nonabelian.

4

Matrices

In mathematics, a matrix (plural matrices) is a rectangular array of numbers, symbols, or expressions, arranged in rows and columns. The individual items (numbers, symbols or expressions) in a matrix are called its elements or entries. Provided that they are the same size (have the same number of rows and the same number of columns) two matrices can be added or subtracted element by element. The rule for matrix multiplication, however, is that two matrices can be multiplied only when the number of columns in the first equals the number of rows in the second. Any matrix can be multiplied element-wise by a scalar from its associated field.

Matrices which have a single row are called row vectors, and those which have a single column are called column vectors. A matrix which has the same number of rows and columns is called a square matrix. In some contexts, such as computer algebra programs, it is useful to consider a matrix with no rows or no columns, called an empty matrix.

For example, the dimension of the matrix below is 2×3 (read "two by three"), because there are two rows and three columns:

$$\begin{bmatrix} 1 & 9 & -13 \\ 20 & 5 & -6 \end{bmatrix}.$$

Provided that they have the same size (each matrix has the same number of rows and the same number of columns as the other), two matrices can be added or subtracted element by element (see conformable matrix). The rule for matrix multiplication, however, is that two matrices can be multiplied only when the number of columns in the first equals the number of rows in the second (i.e., the inner dimensions are the same, n for an $(m\times n)$-matrix times an $(n\times p)$-matrix, resulting in an $(m\times p)$-matrix). There is no product the other way round, a first hint that matrix multiplication is not commutative. Any matrix can be multiplied element-wise by a scalar from its associated field.

The individual items in an m×n matrix A, often denoted by $a_{i,j}$, where i and j usually vary from 1 to m and n, respectively, are called its elements or entries. For conveniently expressing an element of the results of matrix operations the indices of the element are often attached to the parenthesized or bracketed matrix expression; e.g., $(AB)_{i,j}$ refers to an element of a matrix product. In the context of abstract index notation this ambiguously refers also to the whole matrix product.

A major application of matrices is to represent linear transformations, that is, generalizations of linear functions such as f(x) = 4x. For example, the rotation of vectors in three-dimensional space is a linear transformation, which can be represented by a rotation matrix R: if v is a column vector (a matrix with only one column) describing the position of a point in space, the product Rv is a column vector describing the position of that point after a rotation. The product of two transformation matrices is a matrix that represents the composition of two transformations. Another application of matrices is in the solution of systems of linear equations. If the matrix is square, it is possible to deduce some of its properties by computing its determinant. For example, a square matrix has an inverse if and only if its determinant is not zero. Insight into the geometry of a linear transformation is obtainable (along with other information) from the matrix's eigenvalues and eigenvectors.

Applications of matrices are found in most scientific fields. In every branch of physics, including classical mechanics, optics, electromagnetism, quantum mechanics, and quantum electrodynamics, they are used to study physical phenomena, such as the motion of rigid bodies. In computer graphics, they are used to manipulate 3D models and project them onto a 2-dimensional screen. In probability theory and statistics, stochastic matrices are used to describe sets of probabilities; for instance, they are used within the PageRank algorithm that ranks the pages in a Google search. Matrix calculus generalizes classical analytical notions such as derivatives and exponentials to higher dimensions. Matrices are used in economics to describe systems of economic relationships.

A major branch of numerical analysis is devoted to the development of efficient algorithms for matrix computations, a subject that is centuries old and is today an expanding area of research. Matrix decomposition methods simplify computations, both theoretically and practically. Algorithms that are tailored to particular matrix structures, such as sparse matrices and near-diagonal matrices, expedite computations in finite element method and other computations. Infinite matrices occur in planetary theory and in atomic theory. A simple example of an infinite matrix is the matrix representing the derivative operator, which acts on the Taylor series of a function.

Definition of Matrix

A matrix is a rectangular array of numbers or other mathematical objects for which operations such as addition and multiplication are defined. Most commonly, a matrix over a field F is a rectangular array of scalars each of which is a member of F. Most of this article focuses on real and complex matrices, that is, matrices whose elements are real numbers or complex numbers, respectively. More general types of entries are discussed below. For instance, this is a real matrix:

$$\mathbf{A} = \begin{bmatrix} -1.3 & 0.6 \\ 20.4 & 5.5 \\ 9.7 & -6.2 \end{bmatrix}.$$

The numbers, symbols, or expressions in the matrix are called its entries or its elements. The horizontal and vertical lines of entries in a matrix are called rows and columns, respectively.

Size of a matrix

The size of a matrix is defined by the number of rows and columns that it contains. A matrix with m rows and n columns is called an m × n matrix or m-by-n matrix, while m and n are called its dimensions. For example, the matrix A above is a 3 × 2 matrix.

Matrices with a single row are called row vectors, and those with a single column are called column vectors. A matrix with the same number of rows and columns is called a square matrix. A matrix with an infinite number of rows or columns (or both) is called an infinite matrix. In some contexts, such as computer algebra programs, it is useful to consider a matrix with no rows or no columns, called an empty matrix.

Matrices are commonly written in box brackets or parentheses:

$$\mathbf{A} = \begin{bmatrix} a_{11} & a_{12} & \cdots & a_{1n} \\ a_{21} & a_{22} & \cdots & a_{2n} \\ \vdots & \vdots & \ddots & \vdots \\ a_{m1} & a_{m2} & \cdots & a_{mn} \end{bmatrix} = \begin{pmatrix} a_{11} & a_{12} & \cdots & a_{1n} \\ a_{21} & a_{22} & \cdots & a_{2n} \\ \vdots & \vdots & \ddots & \vdots \\ a_{m1} & a_{m2} & \cdots & a_{mn} \end{pmatrix} = (a_{ij}) \in \mathbb{R}^{m \times n}.$$

The specifics of symbolic matrix notation vary widely, with some prevailing trends. Matrices are usually symbolized using upper-case letters (such as A in the examples above), while the corresponding lower-case letters, with two subscript indices (for example, a_{11}, or $a_{1,1}$), represent the entries. In addition to using upper-case letters to symbolize matrices, many authors use a special typographical style, commonly boldface upright (non-italic), to further distinguish matrices from other mathematical objects. An alternative notation involves the use of a double-underline with the variable name, with or without boldface style, (for example, $\underline{\underline{A}}$).

The entry in the i-th row and j-th column of a matrix A is sometimes referred to as the i,j, (i,j), or (i,j)th entry of the matrix, and most commonly denoted as $a_{i,j}$, or a_{ij}. Alternative notations for that entry are A[i,j] or $A_{i,j}$. For example, the (1,3) entry of the following matrix A is 5 (also denoted a_{13}, $a_{1,3}$, A[1,3] or $A_{1,3}$):

$$\mathbf{A} = \begin{bmatrix} 4 & -7 & 5 & 0 \\ 2 & 0 & 11 & 8 \\ 19 & 1 & -3 & 12 \end{bmatrix}$$

Sometimes, the entries of a matrix can be defined by a formula such as $a_{i,j} = f(i, j)$. For example, each of the entries of the following matrix A is determined by $a_{ij} = i$ ″ j.

$$\mathbf{A} = \begin{bmatrix} 0 & -1 & 2 & 3 \\ -1 & 0 & 1 & 2 \\ 2 & 1 & 0 & -1 \end{bmatrix}$$

In this case, the matrix itself is sometimes defined by that formula, within square brackets or double parentheses. For example, the matrix above is defined as A = [i″j], or A = ((i″j)). If matrix size is m × n, the above-mentioned formula f(i, j) is valid for any i = 1, ..., m and any j = 1, ..., n. This can be either specified separately, or using m × n as a subscript. For instance, the matrix A above is 3 × 4 and can be defined as A = [i ″ j] (i = 1, 2, 3; j = 1, ..., 4), or $A = [i \text{ ″ } j]_{3\times4}$.

Some programming languages utilize doubly subscripted arrays (or arrays of arrays) to represent an m-×-n matrix. Some programming languages start the numbering of array indexes at zero, in which case the entries of an m-by-n matrix are indexed by 0 d″ i d″ m ″ 1 and 0 d″ j d″ n ″ 1. This article follows the more common convention in mathematical writing where enumeration starts from 1.

An asterisk is occasionally used to refer to whole rows or columns in a matrix. For example, $a_{i,″}$ refers to the ith row of A, and $a_{″,j}$ refers to the jth column of A. The set of all m-by-n matrices is denoted (m, n).

HISTORY OF THE MATRIX

The matrix has a long history of application in solving linear equations. They were known as arrays until the 1800's. The term "matrix" (Latin for "womb", derived from mater—mother) was coined by James Joseph Sylvester in 1850, who understood a matrix as an object giving rise to a number of determinants today called minors, that is to say, determinants of smaller matrices that are derived from the original one by removing columns and rows. An English mathematician named Cullis was the first to use modern bracket notation for matrices in 1913 and he simultaneously demonstrated the first significant use of the notation A=ai,j to represent a matrix where ai,j refers to the element found in the ith row and the jth column. Matrices can be used to compactly write and work with multiple linear equations, referred to as a system of linear equations, simultaneously. Matrices and matrix multiplication reveal their essential features when related to linear transformations, also known as linear maps.

More uses of matrix-like arrangements of numbers appears in which a method is given for solving simultaneous equations using a counting board that is mathematically identical to the modern matrix method of solution outlined by Carh Fridrich Gauss (1777-1855) also known as Gaussan Elimination.

This project seeks to give an overview of the history of matrices and its practical applications touching on the various topics used in concordance with it. Around 4000 years ago, the people of Babylon knew how to solve a simple2X2 system of linear equations with two unknowns. Around 200 BC, the Chinese published that "Nine Chapters of the Mathematical Art," they displayed the ability to solve a 3X3 system of equations (Perotti). The power and progress in Matrices and its application did not come to fruition until the late 17th century.

The emergence of the subject came from determinants, values connected to a square matrix, studied by the founder of calculus, Leibnitz, in the late 17th century. Lagrange came out with his work regarding Lagrange multipliers, a way to "characterize the maxima and minima multivariate functions." (Darkwing) More than fifty years later, Cramer presented his ideas of solving systems of linear equations based on determinants more than50 years after Leibnitz (Darkwing). Interestingly enough, Cramer provided no proof for solving an n x n system.

As mentioned before, Gauss work dealt much with solving linear equations themselves initially, but did not have as much to do with matrices. In order for matrix algebra to develop, a proper notation-or method of describing the process was necessary. Also vital to this process was a definition of matrix multiplication and the facets involving it. "The introduction of matrix notation and the invention of the word matrix were motivated by attempts to develop the right algebraic language for studying determinants. In 1848, J.J. Sylvester introduced the term "matrix," the Latin word for womb, as a name for an array of numbers. He used womb, because see, linear algebra has become more relevant since the emergence of calculus even though its foundational equation of ax+ b=0 dates back centuries.

Euler brought to light the idea that a system of equations doesn't necessarily have to have a solution. He recognized the need for conditions to be placed upon unknown variables in order to find a solution. The initial work up until this period mainly dealt with the concept of unique solutions and square matrices where the number of equations matched the number of unknowns.

With the turn into the 19th century Gauss introduced a procedure to be used for solving a system of linear equations. His work dealt mainly with the linear equations and had yet to bring in the idea of matrices or their notations. His efforts dealt with equations of differing numbers and variables as well as the traditional pre-19th century works of Euler, Leibnitz, and Cramer. Gauss' work is now summed up in the term Gaussian elimination. This method uses the concepts of combining, swapping, or multiplying rows with each other in order to eliminate variables from certain equations. After variables are determined, the student is then to use back substitution to help find the remaining unknown variables.

Reviewed a matrix as a generator of determinants. The other part, matrix multiplication or matrix algebra came from the work of Arthur Cayley in 1855. Cayley's defined matrix multiplication as, "the matrix of coefficients for the composite transformation T2T1 is the product of the matrix for T2times the matrix of T1". His work dealing with Matrix multiplication culminated in his theorem, the Cayley-Hamilton Theorem. Simply stated, a square matrix satisfies Matrices at the end of the 19th century were heavily connected with Physics issues and for mathematicians, more attention was given to vectors as they proved to be basic mathematical elements. With the advancement of technology using the methods of Cayley, Gauss, Leibnitz, Euler, and others determinants and linear algebra moved forward more quickly and more effective. Regardless of the technology though Gaussian elimination still proves to be the best way known to solve a system of linear equations .

The influence of matrices and it's applications in the mathematical world is spread wide because it provides an important base to many of the principles and practices. Some of the things Matrices is used for are to solve systems of linear format, to find least-square best fit lines to predict future outcomes or find trends, to encode and decode messages. Other more broad topics that it is used for are to solve questions of energy in Quantum mechanics. It is also used to create simple everyday household games like Sudoku. It is because of these practical applications that Matrices has spread so far and advanced. The key, however, is to understand that the history of linear algebra provides the basis for these applications.

Although linear algebra is a fairly new subject when compared to other mathematical practices, its uses are widespread. With the efforts of calculus-savvy Leibnitz the concept of using systems of linear equations to solve unknowns was formalized. Other efforts from Scholars like Cayley. Euler, Sylvester, and others changed matrices into the use of linear algebra to represent them. Gauss brought his theory to solve systems of equations proving to be the most effective basis for solving unknowns.

Technology continues to push the use further and further, but the history of matrices and its application continues to provide the foundation. Even though every few years companies update their textbooks, the fundamentals stay the same.

Basic Operations

There are a number of basic operations that can be applied to modify matrices, called matrix addition, scalar multiplication, transposition, matrix multiplication, row operations, and submatrix.

Types of Matrices

There are various types of matrices, depending on their structure. Let's explore the most common types:

Null Matrix

A matrix that has all 0 elements is called a null matrix. It can be of any order. For example, we could have a null matrix of the order 2 X 3. It's also a singular matrix, since it does not have an inverse and its determinant is 0.

$$\begin{bmatrix} 0 & 0 & 0 \\ 0 & 0 & 0 \end{bmatrix}$$

Any matrix that does have an inverse can be called a regular matrix.

Row Matrix

A row matrix is a matrix with only one row. Its order would be 1 X C, where C is the number of columns. For example, here's a row matrix of the order 1 X 5:

$$\begin{bmatrix} 3 & 5 & 7 & 9 & 11 \end{bmatrix}$$

Column Matrix

A column matrix is a matrix with only one column. It is represented by an order of R X 1, where R is the number of rows. Here's a column matrix of the order 3 X 1:

$$\begin{bmatrix} 2 \\ 5 \\ 7 \end{bmatrix}$$

Square Matrix

A matrix where the number of rows is equal to the number of columns is called a square matrix. Here's a square matrix of the order 2 X 2:

$$\begin{bmatrix} 4 & 9 \\ 15 & 2 \end{bmatrix}$$

Diagonal Matrix

A diagonal matrix is a square matrix where all the elements are 0 except for those in the diagonal from the top left corner to the bottom right corner. Let's take a look at a diagonal matrix of order 4 X 4:

$$\begin{bmatrix} 8 & 0 & 0 & 0 \\ 0 & 5 & 0 & 0 \\ 0 & 0 & 2 & 0 \\ 0 & 0 & 0 & 9 \end{bmatrix}$$

A special type of diagonal matrix, where all the diagonal elements are equal is called a scalar matrix. We can see a 3 X 3 scalar matrix here:

$$\begin{bmatrix} 2 & 0 & 0 \\ 0 & 2 & 0 \\ 0 & 0 & 2 \end{bmatrix}$$

A scalar matrix whose diagonal elements are all 1 is called a unit matrix, or identity matrix.

$$\begin{bmatrix} 1 & 0 & 0 \\ 0 & 1 & 0 \\ 0 & 0 & 1 \end{bmatrix}$$

Upper Triangular Matrix

A square matrix where all the elements below the left-right diagonal are 0 is called an upper triangular matrix. Here's an upper triangular matrix of order 3 X 3:

$$\begin{bmatrix} 2 & 5 & 8 \\ 0 & 6 & 3 \\ 0 & 0 & 9 \end{bmatrix}$$

Lower Triangular Matrix

A square matrix where all the elements above the left-right diagonal are 0 is called a lower triangular matrix. Here's what a lower triangular matrix of order 3 X 3 could look like:

$$\begin{bmatrix} 10 & 0 & 0 \\ 8 & 7 & 0 \\ 3 & 2 & 9 \end{bmatrix}$$

Symmetric Matrix

A matrix whose transpose is the same as the original matrix is called a symmetric matrix. Only a square matrix can be a symmetric matrix. The transpose of a matrix is another matrix that is formed by switching the rows and columns of a given matrix. The given matrix A is a 3 X 3 symmetric matrix, since it's the same as its transpose A^T.

$$A = \begin{bmatrix} 5 & 1 & 9 \\ 1 & 8 & 6 \\ 9 & 6 & 5 \end{bmatrix} ; A^T = \begin{bmatrix} 5 & 1 & 9 \\ 1 & 8 & 6 \\ 9 & 6 & 5 \end{bmatrix}$$

Antisymmetric Matrix

A square matrix whose transpose is its negation is an antisymmetric matrix, or skew-symmetric matrix. The negation of a matrix is a matrix formed by negating the signs of all the entries:

$$A = \begin{bmatrix} 0 & 2 & -1 \\ -2 & 0 & -4 \\ 1 & 4 & 0 \end{bmatrix} , A^T = \begin{bmatrix} 0 & -2 & 1 \\ 2 & 0 & 4 \\ -1 & -4 & 0 \end{bmatrix}$$

Addition, scalar multiplication, and transposition

Matrix addition, subtraction, and scalar multiplication are types of operations that can be applied to modify matrices. There are a number of operations that can be applied to modify matrices, such as matrix addition, subtraction, and scalar multiplication. These form the basic techniques to work with matrices.

These techniques can be used in calculating sums, differences and products of information such as sodas that come in three different flavors: apple, orange, and strawberry and two different packaging: bottle and can. Two tables summarizing the total sales between last month and this month are written to illustrate the amounts. Matrix addition, subtraction and scalar multiplication can be used to find such things as: the sales of last month and the sales of this month, the average sales for each flavor and packaging of soda in the 2-month period.

The sum A+B of two m-by-n matrices A and B is calculated entrywise: $(A + B)_{i,j} = A_{i,j} + B_{i,j}$, where 1 d" i d" m and 1 d" j d" n.

$$\begin{bmatrix} 1 & 3 & 1 \\ 1 & 0 & 0 \end{bmatrix} + \begin{bmatrix} 0 & 0 & 5 \\ 7 & 5 & 0 \end{bmatrix} = \begin{bmatrix} 1+0 & 3+0 & 1+5 \\ 1+7 & 0+5 & 0+0 \end{bmatrix} = \begin{bmatrix} 1 & 3 & 6 \\ 8 & 5 & 0 \end{bmatrix}$$

Scalar multiplication

The product cA of a number c (also called a scalar in the parlance of abstract algebra) and a matrix A is computed by multiplying every entry of A by c:

$(cA)_{i,j} = c \cdot A_{i,j}$.

This operation is called scalar multiplication, but its result is not named "scalar product" to avoid confusion, since "scalar product" is sometimes used as a synonym for "inner product".

$$2\cdot\begin{bmatrix} 1 & 8 & -3 \\ 4 & -2 & 5 \end{bmatrix} = \begin{bmatrix} 2\cdot1 & 2\cdot8 & 2\cdot-3 \\ 2\cdot4 & 2\cdot-2 & 2\cdot5 \end{bmatrix} = \begin{bmatrix} 2 & 16 & -6 \\ 8 & -4 & 10 \end{bmatrix}$$

Transposition

The transpose of an m-by-n matrix A is the n-by-m matrix A^T (also denoted A^{tr} or ${}^tA$) formed by turning rows into columns and vice versa:

$(A^T)_{i,j} = A_{j,i}$.

$$\begin{bmatrix} 1 & 2 & 3 \\ 0 & -6 & 7 \end{bmatrix}^T = \begin{bmatrix} 1 & 0 \\ 2 & -6 \\ 3 & 7 \end{bmatrix}$$

Familiar properties of numbers extend to these operations of matrices: for example, addition is commutative, that is, the matrix sum does not depend on the order of the summands: A + B = B + A. The transpose is compatible with addition and scalar multiplication, as expressed by $(cA)^T = c(A^T)$ and $(A + B)^T = A^T + B^T$. Finally, $(A^T)^T = A$.

Matrix multiplication

Multiplication of two matrices is defined if and only if the number of columns of the left matrix is the same as the number of rows of the right matrix. If A is an m-by-n matrix and B is an n-by-p matrix, then their matrix product AB is the m-by-p matrix whose entries are given by dot product of the corresponding row of A and the corresponding column of B:

$$[\mathbf{AB}]_{i,j} = a_{i,1}b_{1,j} + a_{i,2}b_{2,j} + \cdots + a_{i,n}b_{n,j} = \sum_{r=1}^{n} a_{i,r}b_{r,j},$$

where 1 d″ i d″ m and 1 d″ j d″ p. For example, the underlined entry 2340 in the product is calculated as (2 × 1000) + (3 × 100) + (4 × 10) = 2340:

$$\begin{bmatrix} \underline{2} & \underline{3} & \underline{4} \\ 1 & 0 & 0 \end{bmatrix} \begin{bmatrix} 0 & \underline{1000} \\ 1 & \underline{100} \\ 0 & \underline{10} \end{bmatrix} = \begin{bmatrix} 3 & \underline{2340} \\ 0 & 1000 \end{bmatrix}.$$

Matrix multiplication satisfies the rules (AB)C = A(BC) (associativity), and (A + B)C = AC + BC as well as C(A + B) = CA + CB (left and right distributivity), whenever the size of the matrices is such that the various products are defined. The product AB may be defined without BA being defined, namely if A and B are m-by-n and n-by-k matrices, respectively, and m ‴ k. Even if both products are defined, they need not be equal, that is, generally

AB ‴ BA,

that is, matrix multiplication is not commutative, in marked contrast to (rational, real, or complex) numbers whose product is independent of the order of the factors. An example of two matrices not commuting with each other is:

$$\begin{bmatrix} 1 & 2 \\ 3 & 4 \end{bmatrix} \begin{bmatrix} 0 & 1 \\ 0 & 0 \end{bmatrix} = \begin{bmatrix} 0 & 1 \\ 0 & 3 \end{bmatrix},$$

whereas

$$\begin{bmatrix} 0 & 1 \\ 0 & 0 \end{bmatrix} \begin{bmatrix} 1 & 2 \\ 3 & 4 \end{bmatrix} = \begin{bmatrix} 3 & 4 \\ 0 & 0 \end{bmatrix}.$$

Besides the ordinary matrix multiplication just described, there exist other less frequently used operations on matrices that can be considered forms of multiplication, such as the Hadamard product and the Kronecker product. They arise in solving matrix equations such as the Sylvester equation.

Row operations

There are three types of row operations:

1. row addition, that is adding a row to another.
2. row multiplication, that is multiplying all entries of a row by a non-zero constant;
3. row switching, that is interchanging two rows of a matrix;

These operations are used in a number of ways, including solving linear equations and finding matrix inverses.

Submatrix

A submatrix of a matrix is obtained by deleting any collection of rows and/or columns. For example, from the following 3-by-4 matrix, we can construct a 2-by-3 submatrix by removing row 3 and column 2:

$$\mathbf{A} = \begin{bmatrix} 1 & 2 & 3 & 4 \\ 5 & 6 & 7 & 8 \\ 9 & 10 & 11 & 12 \end{bmatrix} \rightarrow \begin{bmatrix} 1 & 3 & 4 \\ 5 & 7 & 8 \end{bmatrix}.$$

The minors and cofactors of a matrix are found by computing the determinant of certain submatrices.

A principal submatrix is a square submatrix obtained by removing certain rows and columns. The definition varies from author to author. According to some authors, a principal submatrix is a submatrix in which the set of row indices that remain is the same as the set of column indices that remain. Other authors define a principal submatrix as one in which the first k rows and columns, for some number k, are the ones that remain; this type of submatrix has also been called a leading principal submatrix.

Basis transformation matrix

Let B and C be two different bases of a vector space V, and let us mark with $[M]_C^B$ the matrix which has columns consisting of the C representation of basis vectors $b_1, b_2, \ldots, b_n$:

$$[M]_C^B = [[b_1]_C \quad \cdots \quad [b_n]_C]$$

This matrix is referred to as the basis transformation matrix from B to C. It can be regarded as an automorphism over V. Any vector v represented in B can be transformed to a representation in C as follows:

$$[v]_C = [M]_C^B [v]_B.$$

If E is the standard basis, the notation can be simplified by omitting it, with the transformation from B to E being represented:

$$v = [M]^B [v]_B.$$

where

$$v = [v]_E:$$

$$[M]^B = [M]_E^B.$$

Under the transformation of basis, notice that the superscript on the transformation matrix, M, and the subscript on the coordinate vector, v, are the same, and seemingly cancel, leaving the remaining subscript. While this may serve as a memory aid, it is important to note that no such cancellation, or similar mathematical operation, is taking place.

Corollary

The matrix M is an invertible matrix and M^{-1} is the basis transformation matrix from C to B. In other words,

$$\begin{aligned} \mathrm{Id} \\ &= [M]_C^B [M]_B^C = [M]_C^C \\ &= [M]_B^C [M]_C^B = [M]_B^B \end{aligned}$$

Infinite-dimensional vector spaces

Suppose V is an infinite-dimensional vector space over a field F. If the dimension is ê, then there is some basis of ê elements for V. After an order is chosen, the basis can be considered an ordered basis. The elements of V are finite linear combinations of elements in the basis, which give rise to unique coordinate representations exactly as described before. The only change is that the indexing set for the coordinates is not finite. Since a given vector v is a finite linear combination of basis elements, the only nonzero entries of the coordinate vector for v will be the nonzero coefficients of the linear combination representing v. Thus the coordinate vector for v is zero except in finitely many entries.

The linear transformations between (possibly) infinite-dimensional vector spaces can be modeled, analogously to the finite-dimensional case, with infinite matrices. The special case of the transformations from V into V is described in the full linear ring article.

2 × 2 *real matrices*

In mathematics, the associative algebra of 2×2 real matrices is denoted by M(2, R). Two matrices p and q in M(2, R) have a sum p + q given by matrix addition. The product matrix p q is formed from the dot product of the rows and columns of its factors through matrix multiplication. For

$$q = \begin{pmatrix} a & b \\ c & d \end{pmatrix},$$

let

$$q^* = \begin{pmatrix} d & -b \\ -c & a \end{pmatrix}.$$

Then $q\,q^* = q^*q = (ad - bc)\,I$, where I is the 2×2 identity matrix. The real number $ad - bc$ is called the determinant of q. When $ad - bc \neq 0$, q is an invertible matrix, and then

$$q^{-1} = \frac{q^*}{ad - bc}.$$

The collection of all such invertible matrices constitutes the general linear group GL(2, R). In terms of abstract algebra, M(2, R) with the associated addition and multiplication operations forms a ring, and GL(2, R) is its group of units. M(2, R) is also a four-dimensional vector space, so it is considered an associative algebra.

The 2×2 real matrices are in one-one correspondence with the linear mappings of the two-dimensional Cartesian coordinate system into itself by the rule

$$\begin{pmatrix} x \\ y \end{pmatrix} \mapsto \begin{pmatrix} a & b \\ c & d \end{pmatrix}\begin{pmatrix} x \\ y \end{pmatrix} = \begin{pmatrix} ax + by \\ cx + dy \end{pmatrix}.$$

The next section displays M(2,R) is a union of planar cross sections that include a real line. M(2,R) is ring isomorphic to split-quaternions, where there is a similar union but with index sets that are hyperboloids.

Profile

Within M(2, R), the multiples by real numbers of the identity matrix I may be considered a real line. This real line is the place where all commutative subrings come together:

Let $P_m = \{xI + ym : x, y \in R\}$ where $m^2 \in \{-I, 0, I\}$. Then P_m is a commutative subring and $M(2, R) = \bigcup P_m$ where the union is over all m such that $m^2 \in \{-I, 0, I\}$.

To identify such m, first square the generic matrix:

$$\begin{pmatrix} aa + bc & ab + bd \\ ac + cd & bc + dd \end{pmatrix}.$$

When a + d = 0 this square is a diagonal matrix.

Thus one assumes d = ″a when looking for m to form commutative subrings. When mm = ″I, then bc = ″1 ″ aa, an equation describing a hyperbolic paraboloid in the space of parameters (a, b, c). Such an m serves as an imaginary unit. In this case P_m is isomorphic to the field of (ordinary) complex numbers.

When mm = +I, m is an involutory matrix. Then bc = +1 ″ aa, also giving a hyperbolic paraboloid. If a matrix is an idempotent matrix, it must lie in such a P_m and in this case P_m is isomorphic to the ring of split-complex numbers.

The case of a nilpotent matrix, mm = 0, arises when only one of b or c is non-zero, and the commutative subring P_m is then a copy of the dual number plane.

When M(2, R) is reconfigured with a change of basis, this profile changes to the profile of split-quaternions where the sets of square roots of I and "I take a symmetrical shape as hyperboloids.

Equi-areal mapping

First transform one differential vector into another:

$$\begin{pmatrix} du \\ dv \end{pmatrix} = \begin{pmatrix} p & r \\ q & s \end{pmatrix}\begin{pmatrix} dx \\ dy \end{pmatrix} = \begin{pmatrix} p\,dx + r\,dy \\ q\,dx + s\,dy \end{pmatrix}.$$

Areas are measured with density d x ‴ d y {\displaystyle dx\wedge dy}, a differential 2-form which involves the use of exterior algebra. The transformed density is

$$\begin{aligned} du \wedge dv &= 0 + ps\, dx \wedge dy + qr\, dy \wedge dx + 0 \\ &= (ps - qr)\, dx \wedge dy \\ &= \det(g)\, dx \wedge dy. \end{aligned}$$

Thus the equi-areal mappings are identified with SL(2,/ R) = {g ″ M(2,/ R) : det(g) = 1}, the special linear group. Given the profile above, every such g lies in a commutative subring P_m representing a type of complex plane according to the square of m. Since g/ g^* = I, one of the following three alternatives occurs:

- mm = "I and g is on a circle of Euclidean rotations; or
- mm = I and g is on an hyperbola of squeeze mappings; or
- mm = 0 and g is on a line of shear mappings.

Writing about planar affine mapping, Rafael Artzy made a similar trichotomy of planar, linear mapping in his book Linear Geometry (1965).

Functions of 2 × 2 real matrices

The commutative subrings of M(2, R) determine the function theory; in particular the three types of subplanes have their own algebraic structures which set the value of algebraic expressions. Consideration of the square root function and the logarithm function serves to illustrate the constraints implied by the special properties of each type of subplane P_m described in the above profile. The concept of identity component of the group of units of P_m leads to the polar decomposition of elements of the group of units:

- If mm = "I, then z = ñ exp(èm).
- If mm = 0, then z = ñ exp(s m) or z = "ñ exp(s m).
- If mm = I, then z = ñ exp(a m) or z = "ñ exp(a m) or z = m ñ exp(a m) or z = "m ñ exp(a m).

In the first case exp(è m) = cos(è) + m sin(è). In the case of the dual numbers exp(s m) = 1 + s m. Finally, in the case of split complex numbers there are four components in the group of units. The identity component is parameterized by ñ and exp(a m) = cosh(a) + m sinh(a).

Now $\sqrt{\square\exp(am)}$ $\sqrt{\ }$ $\exp\frac{\square 1}{\square 2}am$ regardless of the subplane P_m, but the argument of the function must be taken from the identity component of its group of units. Half the plane is lost in the case of the dual number structure; three-quarters of the plane must be excluded in the case of the split-complex number structure.

LINEAR SYSTEMS OF EQUATIONS WITH AUGMENTED MATRICES

In mathematics, a system of linear equations (or linear system) is a collection of one or more linear equations involving the same set of variables. For example,

$$3x \square 2y \square z \square 1$$

$$2x \square 2y \square 4z \quad \square 2$$

$$\square x \square \tfrac{1}{2} y \square z \square 0$$

is a system of three equations in the three variables x, y, z. A solution to a linear system is an assignment of values to the variables such that all the equations are simultaneously satisfied. A solution to the system above is given by

$$x = 1$$
$$y = -2$$
$$z = -2$$

since it makes all three equations valid. The word "system" indicates that the equations are to be considered collectively, rather than individually.

In mathematics, the theory of linear systems is the basis and a fundamental part of linear algebra, a subject which is used in most parts of modern mathematics. Computational algorithms for finding the solutions are an important part of numerical linear algebra, and play a prominent role in engineering, physics, chemistry, computer science, and economics. A system of non-linear equations can often be approximated by a linear system (see linearization), a helpful technique when making a mathematical model or computer simulation of a relatively complex system.

Very often, the coefficients of the equations are real or complex numbers and the solutions are searched in the same set of numbers, but the theory and the algorithms apply for coefficients and solutions in any field. For solutions in an integral domain like the ring of the integers, or in other algebraic structures, other theories have been developed, see Linear equation over a ring. Integer linear programming is a collection of methods for finding the "best" integer solution (when there are many). Gröbner basis theory provides algorithms when coefficients and unknowns are polynomials. Also tropical geometry is an example of linear algebra in a more exotic structure.

Elementary examples

Trivial example

The system of one equation in one unknown

$$2x = 4$$

has the solution

$$x = 2.$$

However, a linear system is commonly considered as having at least two equations.

Simple nontrivial example

The simplest kind of nontrivial linear system involves two equations and two variables:

$$2x + 3y = 6$$
$$4x + 9y = 15.$$

One method for solving such a system is as follows. First, solve the top equation for x in terms of y:

$$x = 3 - \frac{3}{2}y.$$

Now substitute this expression for x into the bottom equation:

$$4\left(3 - \frac{3}{2}y\right) + 9y = 15.$$

This results in a single equation involving only the variable y. Solving gives y = 1, and substituting this back into the equation for x yields x = 3/ 2 . This method generalizes to systems with additional variables.

General form

A general system of m linear equations with n unknowns can be written as

$$\begin{aligned} a_{11}x_1 + a_{12}x_2 + \cdots + a_{1n}x_n &= b_1 \\ a_{21}x_1 + a_{22}x_2 + \cdots + a_{2n}x_n &= b_2 \\ &\vdots \\ a_{m1}x_1 + a_{m2}x_2 + \cdots + a_{mn}x_n &= b_m, \end{aligned}$$

where $x_1, x_2, \ldots, x_n$ are the unknowns, $a_{11}, a_{12}, \ldots, a_{mn}$ are the coefficients of the system, and $b_1, b_2, \ldots, b_m$ are the constant terms.

Vector equation

One extremely helpful view is that each unknown is a weight for a column vector in a linear combination.

$$x_1 \begin{bmatrix} a_{11} \\ a_{21} \\ \vdots \\ a_{m1} \end{bmatrix} + x_2 \begin{bmatrix} a_{12} \\ a_{22} \\ \vdots \\ a_{m2} \end{bmatrix} + \cdots + x_n \begin{bmatrix} a_{1n} \\ a_{2n} \\ \vdots \\ a_{mn} \end{bmatrix} = \begin{bmatrix} b_1 \\ b_2 \\ \vdots \\ b_m \end{bmatrix}$$

This allows all the language and theory of vector spaces (or more generally, modules) to be brought to bear. For example, the collection of all possible linear combinations of the vectors on the left-hand side is called their span, and the equations have a solution just when the right-hand vector is within that span. If every vector within that span has exactly one expression as a linear combination of the given left-hand vectors, then any solution is unique. In any event, the span has a basis of linearly independent vectors that do guarantee exactly one expression; and the number of vectors in that basis (its dimension) cannot be larger than m or n, but it can be smaller. This is important because if we have m independent vectors a solution is guaranteed regardless of the right-hand side, and otherwise not guaranteed.

Matrix equation

The vector equation is equivalent to a matrix equation of the form

$A\mathbf{x} = \mathbf{b}$

where A is an m×n matrix, x is a column vector with n entries, and b is a column vector with m entries.

$$A = \begin{bmatrix} a_{11} & a_{12} & \cdots & a_{1n} \\ a_{21} & a_{22} & \cdots & a_{2n} \\ \vdots & \vdots & \ddots & \vdots \\ a_{m1} & a_{m2} & \cdots & a_{mn} \end{bmatrix}, \quad \mathbf{x} = \begin{bmatrix} x_1 \\ x_2 \\ \vdots \\ x_n \end{bmatrix}, \quad \mathbf{b} = \begin{bmatrix} b_1 \\ b_2 \\ \vdots \\ b_m \end{bmatrix}$$

The number of vectors in a basis for the span is now expressed as the rank of the matrix.

Solution set

A solution of a linear system is an assignment of values to the variables $x_1, x_2, \ldots, x_n$ such that each of the equations is satisfied. The set of all possible solutions is called the solution set.

The solution set for two equations in three variables is, in general, a line.

A linear system may behave in any one of three possible ways:

1. The system has infinitely many solutions.
2. The system has a single unique solution.
3. The system has no solution.

Geometric interpretation

For a system involving two variables (x and y), each linear equation determines a line on the xy-plane. Because a solution to a linear system must satisfy all of the equations, the solution set is the intersection of these lines, and is hence either a line, a single point, or the empty set.

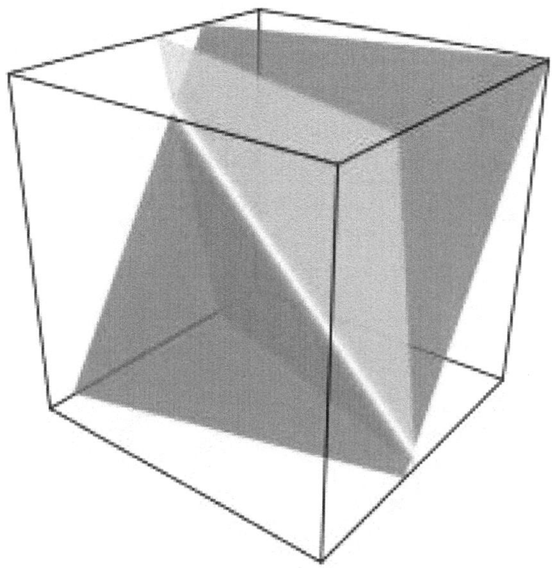

The solution set for two equations in three variables is, in general, a line.

For three variables, each linear equation determines a plane in three-dimensional space, and the solution set is the intersection of these planes. Thus the solution set may be a plane, a line, a single point, or the empty set. For example, as three parallel planes do not have a common point, the solution set of their equations is empty; the solution set of the equations of three planes intersecting at a point is single point; if three planes pass through two points, their equations have at least two common solutions; in fact the solution set is infinite and consists in all the line passing through these points.

For n variables, each linear equation determines a hyperplane in n-dimensional space. The solution set is the intersection of these hyperplanes, and is a flat, which may have any dimension lower than n.

General behavior

The solution set for two equations in three variables is, in general, a line.

In general, the behavior of a linear system is determined by the relationship between the number of equations and the number of unknowns. Here, "in general" means that a different behavior may occur for specific values of the coefficients of the equations.

- In general, a system with fewer equations than unknowns has infinitely many solutions, but it may have no solution. Such a system is known as an underdetermined system.

- In general, a system with the same number of equations and unknowns has a single unique solution.
- In general, a system with more equations than unknowns has no solution. Such a system is also known as an overdetermined system.

In the first case, the dimension of the solution set is, in general, equal to n " m, where n is the number of variables and m is the number of equations.

The following pictures illustrate this trichotomy in the case of two variables:

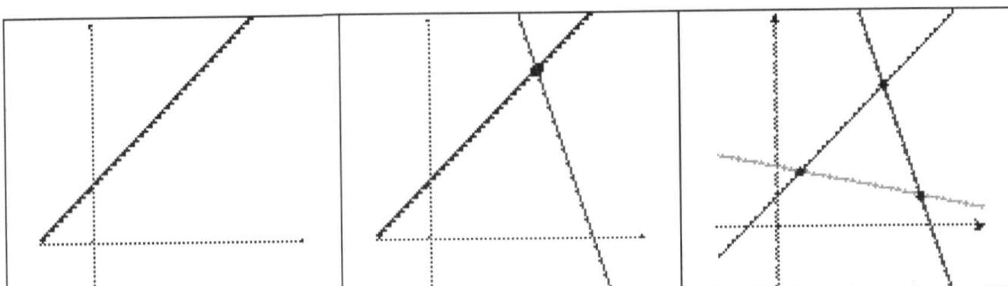

One equation	Two equations	Three equations

The first system has infinitely many solutions, namely all of the points on the blue line. The second system has a single unique solution, namely the intersection of the two lines. The third system has no solutions, since the three lines share no common point.

It must be kept in mind that the pictures above show only the most common case (the general case). It is possible for a system of two equations and two unknowns to have no solution (if the two lines are parallel), or for a system of three equations and two unknowns to be solvable (if the three lines intersect at a single point).

A system of linear equations behave differently from the general case if the equations are linearly dependent, or if it is inconsistent and has no more equations than unknowns.

Properties

Independence

The equations of a linear system are independent if none of the equations can be derived algebraically from the others. When the equations are independent, each equation contains new information about the variables, and removing any of the equations increases the size of the solution set. For linear equations, logical independence is the same as linear independence.

The equations x " 2y = "1, 3x + 5y = 8, and 4x + 3y = 7 are linearly dependent.

For example, the equations

3x □2y □6 and 6x □4y □12

are not independent — they are the same equation when scaled by a factor of two, and they would produce identical graphs. This is an example of equivalence in a system of linear equations.

For a more complicated example, the equations

x □2y □ □1

3x □5y □8

4x □3y □7

are not independent, because the third equation is the sum of the other two. Indeed, any one of these equations can be derived from the other two, and any one of the equations can be removed without affecting the solution set. The graphs of these equations are three lines that intersect at a single point.

Consistency

The equations $3x + 2y = 6$ and $3x + 2y = 12$ are inconsistent.

A linear system is inconsistent if it has no solution, and otherwise it is said to be consistent. When the system is inconsistent, it is possible to derive a contradiction from the equations, that may always be rewritten as the statement $0 = 1$.

For example, the equations

3x □2y □6 and 3x □2y □12

are inconsistent. In fact, by subtracting the first equation from the second one and multiplying both sides of the result by 1/6, we get $0 = 1$. The graphs of these equations on the xy-plane are a pair of parallel lines.

It is possible for three linear equations to be inconsistent, even though any two of them are consistent together. For example, the equations

x □y □1

2x □y □1

3x □2y □3

are inconsistent. Adding the first two equations together gives $3x + 2y = 2$, which can be subtracted from the third equation to yield $0 = 1$. Any two of these equations have a common solution. The same phenomenon can occur for any number of equations.

In general, inconsistencies occur if the left-hand sides of the equations in a system are linearly dependent, and the constant terms do not satisfy the dependence relation. A system of equations whose left-hand sides are linearly independent is always consistent.

Putting it another way, according to the Rouché–Capelli theorem, any system of equations (overdetermined or otherwise) is inconsistent if the rank of the augmented matrix is greater than the rank of the coefficient matrix. If, on the other hand, the ranks of these two matrices are equal, the system must have at least one solution. The solution is unique if and only if the rank equals the number of variables. Otherwise the general solution has k free parameters where k is the difference between the number of variables and the rank; hence in such a case there are an infinitude of solutions. The rank of a system of equations (i.e. the rank of the augmented matrix) can never be higher than [the number of variables] + 1, which means that a system with any number of equations can always be reduced to a system that has a number of independent equations that is at most equal to [the number of variables] + 1.

Equivalence

Two linear systems using the same set of variables are equivalent if each of the equations in the second system can be derived algebraically from the equations in the first system, and vice versa. Two systems are equivalent if either both are inconsistent or each equation of each of them is a linear combination of the equations of the other one. It follows that two linear systems are equivalent if and only if they have the same solution set.

Solving a linear system

There are several algorithms for solving a system of linear equations.

Describing the solution

When the solution set is finite, it is reduced to a single element. In this case, the unique solution is described by a sequence of equations whose left-hand sides are the names of the unknowns and right-hand sides are the corresponding values, for example (x □3, y □2, z □6). When an order on the unknowns has been fixed, for example the alphabetical order the solution may be described as a vector of values, like (3 , " 2 , 6) for the previous example.

To describe a set with an infinite number of solutions, typically some of the variables are designated as free (or independent, or as parameters), meaning that they are allowed to take any value, while the remaining variables are dependent on the values of the free variables.

For example, consider the following system:

$$x \square 3y \square 2z \square 5$$

$$3x \square 5y \quad 6z \square 7$$

The solution set to this system can be described by the following equations:

$$x \square \square 7z \quad 1\square \quad \text{and} \quad y \square 3z \square 2.$$

Here z is the free variable, while x and y are dependent on z. Any point in the solution set can be obtained by first choosing a value for z, and then computing the corresponding values for x and y.

Each free variable gives the solution space one degree of freedom, the number of which is equal to the dimension of the solution set. For example, the solution set for the above equation is a line, since a point in the solution set can be chosen by specifying the value of the parameter z. An infinite solution of higher order may describe a plane, or higher-dimensional set.

Different choices for the free variables may lead to different descriptions of the same solution set. For example, the solution to the above equations can alternatively be described as follows:

$$y \square \square \frac{3}{7}x \square \frac{11}{7} \quad \text{and} \quad z \square \square \frac{1}{7}x \quad \frac{1}{7}.$$

Here x is the free variable, and y and z are dependent.

Elimination of variables

The simplest method for solving a system of linear equations is to repeatedly eliminate variables. This method can be described as follows:

1. In the first equation, solve for one of the variables in terms of the others.
2. Substitute this expression into the remaining equations. This yields a system of equations with one fewer equation and one fewer unknown.
3. Repeat until the system is reduced to a single linear equation.
4. Solve this equation, and then back-substitute until the entire solution is found.

For example, consider the following system:

$$x \square 3y \square 2z \square 5$$

$$3x \square 5y \quad 6z \square 7$$

$$2x \square 4y \quad 3z \square 8$$

Solving the first equation for x gives x = 5 + 2z " 3y, and plugging this into the second and third equation yields

□4y □12z □ □8

□2y □7z □ □2

Solving the first of these equations for y yields y = 2 + 3z, and plugging this into the second equation yields z = 2. We now have:

x □5 □2z □3y

y □2 □3z

z □2

Substituting z = 2 into the second equation gives y = 8, and substituting z = 2 and y = 8 into the first equation yields x = “15. Therefore, the solution set is the single point (x, y, z) = (“15, 8, 2).

Other methods

While systems of three or four equations can be readily solved by hand (see Cracovian), computers are often used for larger systems. The standard algorithm for solving a system of linear equations is based on Gaussian elimination with some modifications. Firstly, it is essential to avoid division by small numbers, which may lead to inaccurate results. This can be done by reordering the equations if necessary, a process known as pivoting. Secondly, the algorithm does not exactly do Gaussian elimination, but it computes the LU decomposition of the matrix A. This is mostly an organizational tool, but it is much quicker if one has to solve several systems with the same matrix A but different vectors b.

If the matrix A has some special structure, this can be exploited to obtain faster or more accurate algorithms. For instance, systems with a symmetric positive definite matrix can be solved twice as fast with the Cholesky decomposition. Levinson recursion is a fast method for Toeplitz matrices. Special methods exist also for matrices with many zero elements (so-called sparse matrices), which appear often in applications.

A completely different approach is often taken for very large systems, which would otherwise take too much time or memory. The idea is to start with an initial approximation to the solution (which does not have to be accurate at all), and to change this approximation in several steps to bring it closer to the true solution. Once the approximation is sufficiently accurate, this is taken to be the solution to the system. This leads to the class of iterative methods.

AUGMENTED MATRIX

In linear algebra, an augmented matrix is a matrix obtained by appending the columns of two given matrices, usually for the purpose of performing the same elementary row operations on each of the given matrices.

Given the matrices A and B, where

$$A = \begin{bmatrix} 1 & 3 & 2 \\ 2 & 0 & 1 \\ 5 & 2 & 2 \end{bmatrix}, \quad B = \begin{bmatrix} 4 \\ 3 \\ 1 \end{bmatrix},$$

the augmented matrix (A | B) is written as

$$(A \mid B) = \left[\begin{array}{ccc|c} 1 & 3 & 2 & 4 \\ 2 & 0 & 1 & 3 \\ 5 & 2 & 2 & 1 \end{array}\right].$$

This is useful when solving systems of linear equations.

For a given number of unknowns, the number of solutions to a system of linear equations depends only on the rank of the matrix representing the system and the rank of the corresponding augmented matrix. Specifically, according to the Rouché–Capelli theorem, any system of linear equations is inconsistent (has no solutions) if the rank of the augmented matrix is greater than the rank of the coefficient matrix; if, on the other hand, the ranks of these two matrices are equal, the system must have at least one solution. The solution is unique if and only if the rank equals the number of variables. Otherwise the general solution has k free parameters where k is the difference between the number of variables and the rank; hence in such a case there are an infinitude of solutions.

An augmented matrix may also be used to find the inverse of a matrix by combining it with the identity matrix.

To find the inverse of a matrix

Let C be the square 2×2 matrix

$$C = \begin{bmatrix} 1 & 3 \\ -5 & 0 \end{bmatrix}.$$

To find the inverse of C we create (C | I) where I is the 2×2 identity matrix. We then reduce the part of (C | I) corresponding to C to the identity matrix using only elementary row operations on (C | I).

$$(C \mid I) = \left[\begin{array}{cc|cc} 1 & 3 & 1 & 0 \\ -5 & 0 & 0 & 1 \end{array}\right]$$

$$(I\,|\,C^{-1}) = \begin{bmatrix} 1 & 0 & 0 & -\frac{1}{5} \\ 0 & 1 & \frac{1}{3} & \frac{1}{15} \end{bmatrix}$$

the right part of which is the inverse of the original matrix.

Solution of a linear system

As used in linear algebra, an augmented matrix is used to represent the coefficients and the solution vector of each equation set. For the set of equations

$$x + 2y + 3z = 0$$
$$3x + 4y + 7z = 2$$
$$6x + 5y + 9z = 11$$

the coefficients and constant terms give the matrices

$$A = \begin{bmatrix} 1 & 2 & 3 \\ 3 & 4 & 7 \\ 6 & 5 & 9 \end{bmatrix}, \quad B = \begin{bmatrix} 0 \\ 2 \\ 11 \end{bmatrix},$$

and hence give the augmented matrix

$$(A\,|\,B) = \begin{bmatrix} 1 & 2 & 3 & 0 \\ 3 & 4 & 7 & 2 \\ 6 & 5 & 9 & 11 \end{bmatrix}$$

Note that the rank of the coefficient matrix, which is 3, equals the rank of the augmented matrix, so at least one solution exists; and since this rank equals the number of unknowns, there is exactly one solution.

To obtain the solution, row operations can be performed on the augmented matrix to obtain the identity matrix on the left side, yielding

$$\begin{bmatrix} 1 & 0 & 0 & 4 \\ 0 & 1 & 0 & 1 \\ 0 & 0 & 1 & -2 \end{bmatrix},$$

so the solution of the system is (x, y, z) = (4, 1, -2).

Gaussian elimination

Gaussian elimination, also known as row reduction, is an algorithm in linear algebra for solving a system of linear equations. It is usually understood as a sequence of operations performed on the corresponding matrix of coefficients. This method can also be used to find the rank of a matrix, to calculate the determinant of a matrix, and to calculate the inverse of an invertible square matrix. The method is named after Carl Friedrich Gauss (1777–1855). Some special cases of the method - albeit presented without proof - were known to Chinese mathematicians as early as circa 179 CE.

To perform row reduction on a matrix, one uses a sequence of elementary row operations to modify the matrix until the lower left-hand corner of the matrix is filled with zeros, as much as possible. There are three types of elementary row operations:

- Swapping two rows,
- Multiplying a row by a nonzero number,
- Adding a multiple of one row to another row.

Using these operations, a matrix can always be transformed into an upper triangular matrix, and in fact one that is in row echelon form. Once all of the leading coefficients (the leftmost nonzero entry in each row) are 1, and every column containing a leading coefficient has zeros elsewhere, the matrix is said to be in reduced row echelon form. This final form is unique; in other words, it is independent of the sequence of row operations used. For example, in the following sequence of row operations (where multiple elementary operations might be done at each step), the third and fourth matrices are the ones in row echelon form, and the final matrix is the unique reduced row echelon form.

$$\begin{bmatrix} 1 & 3 & 1 & 9 \\ 1 & 1 & -1 & 1 \\ 3 & 11 & 5 & 35 \end{bmatrix} \to \begin{bmatrix} 1 & 3 & 1 & 9 \\ 0 & -2 & -2 & -8 \\ 0 & 2 & 2 & 8 \end{bmatrix} \to \begin{bmatrix} 1 & 3 & 1 & 9 \\ 0 & -2 & -2 & -8 \\ 0 & 0 & 0 & 0 \end{bmatrix} \to \begin{bmatrix} 1 & 0 & -2 & -3 \\ 0 & 1 & 1 & 4 \\ 0 & 0 & 0 & 0 \end{bmatrix}$$

Using row operations to convert a matrix into reduced row echelon form is sometimes called Gauss–Jordan elimination. Some authors use the term Gaussian elimination to refer to the process until it has reached its upper triangular, or (unreduced) row echelon form. For computational reasons, when solving systems of linear equations, it is sometimes preferable to stop row operations before the matrix is completely reduced.

Definitions and example of algorithm

The process of row reduction makes use of elementary row operations, and can be divided into two parts. The first part (sometimes called forward elimination) reduces a given system to row echelon form, from which one can tell whether there are no solutions, a unique solution, or infinitely many solutions. The second part (sometimes called back substitution) continues to use row operations until the solution is found; in other words, it puts the matrix into reduced row echelon form.

Another point of view, which turns out to be very useful to analyze the algorithm, is that row reduction produces a matrix decomposition of the original matrix. The elementary row operations may be viewed as the multiplication on the left of the original matrix by elementary matrices. Alternatively, a sequence of elementary operations that reduces a single row may be viewed as multiplication by a Frobenius matrix. Then the first part of the algorithm computes an LU decomposition, while the second part writes the original matrix as the product of a uniquely determined invertible matrix and a uniquely determined reduced row echelon matrix.

5

Differential Calculus

In mathematics, differential calculus is a subfield of calculus concerned with the study of the rates at which quantities change. It is one of the two traditional divisions of calculus, the other being integral calculus, the study of the area beneath a curve.

The primary objects of study in differential calculus are the derivative of a function, related notions such as the differential, and their applications. The derivative of a function at a chosen input value describes the rate of change of the function near that input value. The process of finding a derivative is called differentiation. Geometrically, the derivative at a point is the slope of the tangent line to the graph of the function at that point, provided that the derivative exists and is defined at that point. For a real-valued function of a single real variable, the derivative of a function at a point generally determines the best linear approximation to the function at that point.

Differential calculus and integral calculus are connected by the fundamental theorem of calculus, which states that differentiation is the reverse process to integration.

Differentiation has applications to nearly all quantitative disciplines. For example, in physics, the derivative of the displacement of a moving body with respect to time is the velocity of the body, and the derivative of velocity with respect to time is acceleration. The derivative of the momentumof a body equals the force applied to the body; rearranging this derivative statement leads to the famous F = ma equation associated with Newton's second law of motion. The reaction rate of a chemical reaction is a derivative. In operations research, derivatives determine the most efficient ways to transport materials and design factories.

Derivatives are frequently used to find the maxima and minima of a function. Equations involving derivatives are called differential equations and are fundamental in describing natural phenomena. Derivatives and their generalizations appear in many fields of mathematics, such as complex analysis, functional analysis, differential geometry, measure theory, and abstract algebra.

HISTORY OF DIFFERENTIATION

The concept of a derivative in the sense of a tangent line is a very old one, familiar to Greek geometers such as Euclid (c. 300 BC), Archimedes (c. 287–212 BC) and Apollonius of Perga (c. 262–190 BC). Archimedes also introduced the use of infinitesimals, although these were primarily used to study areas and volumes rather than derivatives and tangents; see Archimedes' use of infinitesimals.

The use of infinitesimals to study rates of change can be found in Indian mathematics, perhaps as early as 500 AD, when the astronomer and mathematician Aryabhata (476–550) used infinitesimals to study the orbit of the Moon. The use of infinitesimals to compute rates of change was developed significantly by Bhâskara II (1114–1185); indeed, it has been argued that many of the key notions of differential calculus can be found in his work, such as "Rolle's theorem".

The Persian mathematician, Sharaf al-Dîn al-Tûsî (1135–1213), in his Treatise on Equations, established conditions for some cubic equations to have solutions, by finding the maxima of appropriate cubic polynomials. He proved, for example, that the maximum of the cubic $a x^2 — x^3$ occurs when $x = 2a/3$, and concluded therefrom that the equation $a x^2 — x^3 = c$ has exactly one positive solution when $c = 4 a^3/27$, and two positive solutions whenever $0 < c < 4 a^3/27$. The historian of science, Roshdi Rashed, has argued that al-Tûsî must have used the derivative of the cubic to obtain this result. Rashed's conclusion has been contested by other scholars, however, who argue that he could have obtained the result by other methods which do not require the derivative of the function to be known.

The modern development of calculus is usually credited to Isaac Newton (1643–1727) and Gottfried Wilhelm Leibniz (1646–1716), who provided independent and unified approaches to differentiation and derivatives. The key insight, however, that earned them this credit, was the fundamental theorem of calculus relating differentiation and integration: this rendered obsolete most previous methods for computing areas and volumes, which had not been significantly extended since the time of Ibn al-Haytham (Alhazen). For their ideas on derivatives, both Newton and Leibniz built on significant earlier work by mathematicians such as Pierre de Fermat (1607-1665), Isaac Barrow (1630–1677), René Descartes (1596–1650), Christiaan Huygens (1629–1695), Blaise Pascal (1623–1662) and John Wallis

(1616–1703). Regarding Fermat's influence, Newton once wrote in a letter that "I had the hint of this method [of fluxions] from Fermat's way of drawing tangents, and by applying it to abstract equations, directly and invertedly, I made it general." Isaac Barrow is generally given credit for the early development of the derivative. Nevertheless, Newton and Leibniz remain key figures in the history of differentiation, not least because Newton was the first to apply differentiation to theoretical physics, while Leibniz systematically developed much of the notation still used today.

Since the 17th century many mathematicians have contributed to the theory of differentiation. In the 19th century, calculus was put on a much more rigorous footing by mathematicians such as Augustin Louis Cauchy (1789–1857), Bernhard Riemann (1826–1866), and Karl Weierstrass (1815–1897). It was also during this period that the differentiation was generalized to Euclidean space and the complex plane.

Applications of derivatives

Optimization

If f is a differentiable function on "! (or an open interval) and x is a local maximum or a local minimum of f, then the derivative of f at x is zero. Points where $f'(x) = 0$ are called critical points or stationary points (and the value of f at x is called a critical value). If f is not assumed to be everywhere differentiable, then points at which it fails to be differentiable are also designated critical points.

If f is twice differentiable, then conversely, a critical point x of f can be analysed by considering the second derivative of f at x :

- if it is positive, x is a local minimum;
- if it is negative, x is a local maximum;
- if it is zero, then x could be a local minimum, a local maximum, or neither. (For example, $f(x) = x^3$ has a critical point at $x = 0$, but it has neither a maximum nor a minimum there, whereas $f(x) = \pm x^4$ has a critical point at $x = 0$ and a minimum and a maximum, respectively, there.)

This is called the second derivative test. An alternative approach, called the first derivative test, involves considering the sign of the f′ on each side of the critical point.

Taking derivatives and solving for critical points is therefore often a simple way to find local minima or maxima, which can be useful in optimization. By the extreme value theorem, a continuous function on a closed interval must attain its minimum and maximum values at least once. If the function is differentiable, the minima and maxima can only occur at critical points or endpoints.

This also has applications in graph sketching: once the local minima and maxima of a differentiable function have been found, a rough plot of the graph can be obtained from the observation that it will be either increasing or decreasing between critical points.

In higher dimensions, a critical point of a scalar valued function is a point at which the gradient is zero. The second derivative test can still be used to analyse critical points by considering the eigenvalues of the Hessian matrix of second partial derivatives of the function at the critical point. If all of the eigenvalues are positive, then the point is a local minimum; if all are negative, it is a local maximum. If there are some positive and some negative eigenvalues, then the critical point is called a "saddle point", and if none of these cases hold (i.e., some of the eigenvalues are zero) then the test is considered to be inconclusive.

Calculus of variations

One example of an optimization problem is: Find the shortest curve between two points on a surface, assuming that the curve must also lie on the surface. If the surface is a plane, then the shortest curve is a line. But if the surface is, for example, egg-shaped, then the shortest path is not immediately clear. These paths are called geodesics, and one of the simplest problems in the calculus of variations is finding geodesics. Another example is: Find the smallest area surface filling in a closed curve in space. This surface is called a minimal surface and it, too, can be found using the calculus of variations.

Physics

Calculus is of vital importance in physics: many physical processes are described by equations involving derivatives, called differential equations. Physics is particularly concerned with the way quantities change and develop over time, and the concept of the "time derivative" — the rate of change over time — is essential for the precise definition of several important concepts. In particular, the time derivatives of an object's position are significant in Newtonian physics:

- velocity is the derivative (with respect to time) of an object's displacement (distance from the original position)
- acceleration is the derivative (with respect to time) of an object's velocity, that is, the second derivative (with respect to time) of an object's position.

Derivative

The derivative of a function of a real variable measures the sensitivity to change of the function value (output value) with respect to a change in its argument (input value). Derivatives are a fundamental tool of calculus.

For example, the derivative of the position of a moving object with respect to time is the object's velocity: this measures how quickly the position of the object changes when time advances.

The derivative of a function of a single variable at a chosen input value, when it exists, is the slope of the tangent line to the graph of the function at that point. The tangent line is the best linear approximation of the function near that input value. For this reason, the derivative is often described as the "instantaneous rate of change", the ratio of the instantaneous change in the dependent variable to that of the independent variable.

Derivatives may be generalized to functions of several real variables. In this generalization, the derivative is reinterpreted as a linear transformation whose graph is (after an appropriate translation) the best linear approximation to the graph of the original function. The Jacobian matrix is the matrix that represents this linear transformation with respect to the basis given by the choice of independent and dependent variables. It can be calculated in terms of the partial derivatives with respect to the independent variables. For a real-valued function of several variables, the Jacobian matrix reduces to the gradient vector.

The process of finding a derivative is called differentiation. The reverse process is called antidifferentiation. The fundamental theorem of calculus states that antidifferentiation is the same as integration. Differentiation and integration constitute the two fundamental operations in single-variable calculus.

CONTINUITY AND DIFFERENTIABILITY

If f is differentiable at a, then f must also be continuous at a. As an example, choose a point a and let f be the step function that returns the value 1 for all x less than a, and returns a different value 10 for all x greater than or equal to a. f cannot have a derivative at a. If h is negative, then a + h is on the low part of the step, so the secant line from a to a + h is very steep, and as h tends to zero the slope tends to infinity. If h is positive, then a + h is on the high part of the step, so the secant line from a to a + h has slope zero. Consequently, the secant lines do not approach any single slope, so the limit of the difference quotient does not exist.[Note 2]

However, even if a function is continuous at a point, it may not be differentiable there. For example, the absolute value function given by f(x) = |x| is continuous at x = 0, but it is not differentiable there. If h is positive, then the slope of the secant line from 0 to h is one, whereas if h is negative, then the slope of the secant line from 0 to h is negative one. This can be seen graphically as a "kink" or a "cusp" in the graph at x = 0. Even a function with a smooth graph is not differentiable at a point where its tangent is vertical: For instance, the function given by $f(x) = x^{1/3}$ is not differentiable at $x = 0$.

In summary: for a function f to have a derivative it is necessary for the function f to be continuous, but continuity alone is not sufficient.

Most functions that occur in practice have derivatives at all points or at almost every point. Early in the history of calculus, many mathematicians assumed that a continuous function was differentiable at most points. Under mild conditions, for example if the function is a monotone function or a Lipschitz function, this is true. However, in 1872 Weierstrass found the first example of a function that is continuous everywhere but differentiable nowhere. This example is now known as the Weierstrass function. In 1931, Stefan Banach proved that the set of functions that have a derivative at some point is a meager set in the space of all continuous functions. Informally, this means that hardly do any random continuous functions have a derivative at even one point.

The derivative as a function

Let f be a function that has a derivative at every point in its domain. We can then define a function that maps every point to the value of the derivative of. This function is written f2 and is called the derivative function or the derivative of f.

Sometimes f has a derivative at most, but not all, points of its domain. The function whose value at a equals f2 (a) whenever f2 (a) is defined and elsewhere is undefined is also called the derivative of f. It is still a function, but its domain is strictly smaller than the domain of f.

Using this idea, differentiation becomes a function of functions: The derivative is an operator whose domain is the set of all functions that have derivatives at every point of their domain and whose range is a set of functions. If we denote this operator by D, then D(f) is the function f2 . Since D(f) is a function, it can be evaluated at a point a. By the definition of the derivative function, D(f)(a) = f2 (a).

Higher derivatives

Let f be a differentiable function, and let f 2 be its derivative. The derivative of f 2 (if it has one) is written f 2 2 and is called the second derivative of f. Similarly, the derivative of the second derivative, if it exists, is written f 2 2 2 and is called the third derivative of f. Continuing this process, one can define, if it exists, the nth derivative as the derivative of the (n-1)th derivative. These repeated derivatives are called higher-order derivatives. The nth derivative is also called the derivative of order n. If x(t) represents the position of an object at time t, then the higher-order derivatives of x have physical interpretations. The second derivative of x is the derivative of x2 , the velocity, and by definition this is the object's acceleration. The third derivative of x is defined to be the jerk, and the fourth derivative is defined to be the jounce.

Generalizations

The concept of a derivative can be extended to many other settings. The common thread is that the derivative of a function at a point serves as a linear approximation of the function at that point.

- An important generalization of the derivative concerns complex functions of complex variables, such as functions from (a domain in) the complex numbers C to C. The notion of the derivative of such a function is obtained by replacing real variables with complex variables in the definition. If C is identified with R^2 by writing a complex number z as x + iy, then a differentiable function from C to C is certainly differentiable as a function from R^2 to R^2 (in the sense that its partial derivatives all exist), but the converse is not true in general: the complex derivative only exists if the real derivative is complex linear and this imposes relations between the partial derivatives called the Cauchy–Riemann equations– see holomorphic functions.
- Another generalization concerns functions between differentiable or smooth manifolds. Intuitively speaking such a manifold M is a space that can be approximated near each point x by a vector space called its tangent space: the prototypical example is a smooth surface in R^3. The derivative (or differential) of a (differentiable) map f: M→N between manifolds, at a point x in M, is then a linear map from the tangent space of M at x to the tangent space of N at f(x). The derivative function becomes a map between the tangent bundles of M and N. This definition is fundamental in differential geometry and has many uses – see pushforward (differential) and pullback (differential geometry).
- Differentiation can also be defined for maps between infinite dimensional vector spaces such as Banach spaces and Fréchet spaces. There is a generalization both of the directional derivative, called the Gâteaux derivative, and of the differential, called the Fréchet derivative.
- One deficiency of the classical derivative is that not very many functions are differentiable. Nevertheless, there is a way of extending the notion of the derivative so that all continuous functions and many other functions can be differentiated using a concept known as the weak derivative. The idea is to embed the continuous functions in a larger space called the space of distributions and only require that a function is differentiable "on average".

- The properties of the derivative have inspired the introduction and study of many similar objects in algebra and topology — see, for example, differential algebra.
- The discrete equivalent of differentiation is finite differences. The study of differential calculus is unified with the calculus of finite differences in time scale calculus.
- Also see arithmetic derivative.

Hyperbolic Functions

If you work in engineering or physics, you have to work with a lot of models. Well, to solve modeling problems you'll need hyperbolic functions.

Certain even and odd combinations of the exponential functions e^x and e^{-x} arise so frequently in engineering, physics, and mathematics that they're given special names. Hyperbolic functions are analogous trigonometric functions in that they are named the same as trigonometric functions with the letter 'h' appended to each name.

These special functions are have the same relationship to the hyperbola that trigonometric functions have to the circle. For this reason, they are collectively known as hyperbolic functions and are individually called hyperbolic sine, hyperbolic cosine, and so on. In addition to modeling, they can be used as solutions to some types of partial differential equations.

Think of taking a rope, fixing the two ends, and then letting it hang under the force of gravity. The shape of the rope will naturally form a hyperbolic cosine curve.

Types & Graphs

Hyperbolic Sine: $y = \sinh(x)$

This math statement is read as 'y equals hyperbolic sine x' or 'sinch x.'

The exponential formula is:

$$\sinh(x) = \frac{e^x - e^{-x}}{2}$$

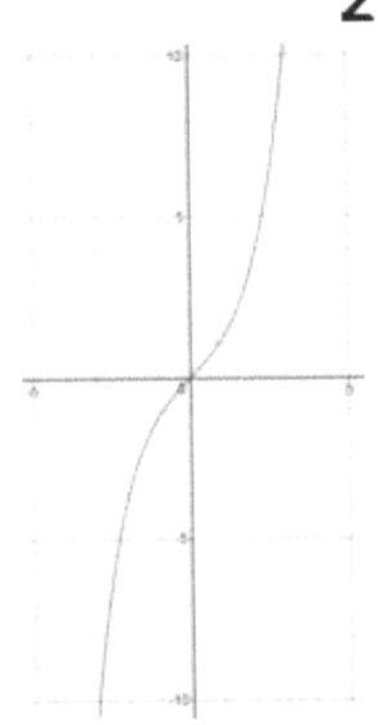

The graph of sinh

Note that the domain and range of this function are both the set of real numbers. This function is also one-to-one.

Hyperbolic Cosine: y = cosh(x)

This math statement is read as 'y equals hyperbolic cosine x' or 'cosh x.'

$$\cosh(x) = \frac{e^x + e^{-x}}{2}$$

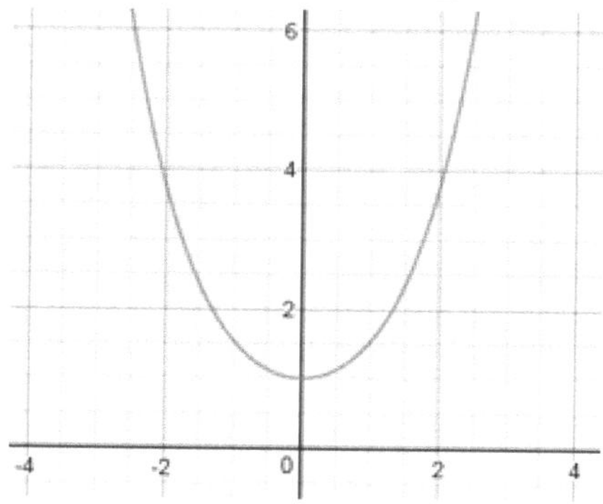

The graph of cosh(x)

Note that the graph is symmetric with respect to the y-axis and it has a minimum vertex at (0,1). The domain of this function is the set of real numbers and the range is any number equal to or greater than one.

Hyperbolic Tangent: y = tanh(x)

This math statement is read as 'y equals hyperbolic tangent x' or 'tanch x.'

$$\tanh(x) = \frac{e^x - e^{-x}}{e^x + e^{-x}} = \frac{\sinh(x)}{\cosh(x)}$$

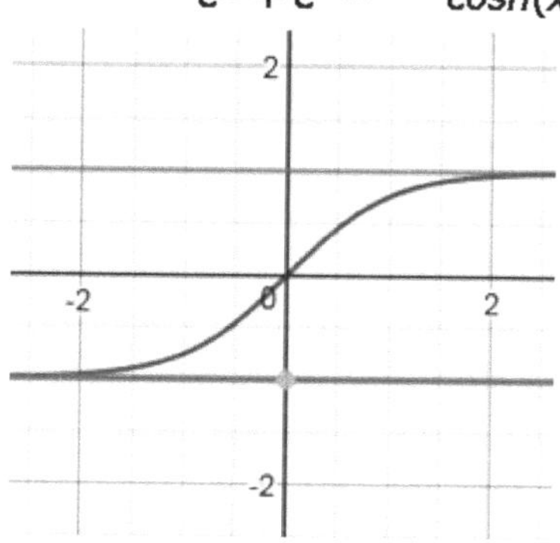

The graph of tanh(x)

Note that the graph is symmetric with respect to the origin. The lines y = 1 and y = -1 are both horizontal asymptotes. The domain of the function is the set of real numbers. The range is the open interval (-1,1).

Hyperbolic Cotangent: y = coth(x)

This math statement is read as 'y equals hyperbolic cotangent x' or 'coth x.'

$$\coth(x) = \frac{e^{x} + e^{-x}}{e^{x} - e^{-x}} = \frac{\cosh(x)}{\sinh(x)}$$

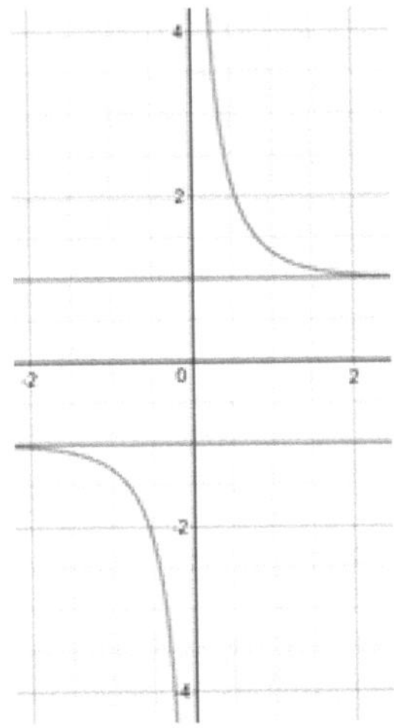

The graph of coth(x)

Note that the graph is symmetric with respect to the origin. The lines y = 1 and y = -1 are both horizontal asymptotes. The y-axis is a vertical asymptote. The domain of the function is the set of real numbers excluding zero. The range is (-infinity,-1) union (1,infinity).

Hyperbolic Secant: y = sech(x)

This math statement is read as 'y equals hyperbolic secant x.'

$$\operatorname{sech}(x) = \frac{2}{e^{x} + e^{-x}} = \frac{1}{\cosh(x)}$$

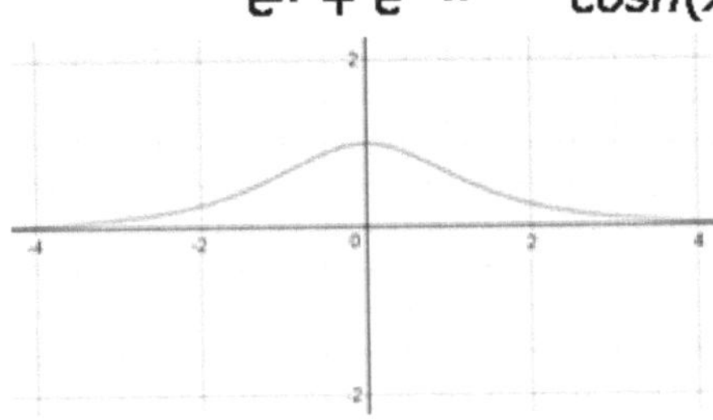

The graph of sech(x)

The x-axis is a horizontal asymptote. The graph is symmetric with respect to the y-axis. The domain of the function is the set of real numbers. The range of the function is the half-open interval (0,1].

Hyperbolic Cosecant: y = csch(x)

This math statement is read as 'y equals hyperbolic cosecant x.'

$$csch(x) = \frac{2}{e^x - e^{-x}} = \frac{1}{sinh(x)}$$

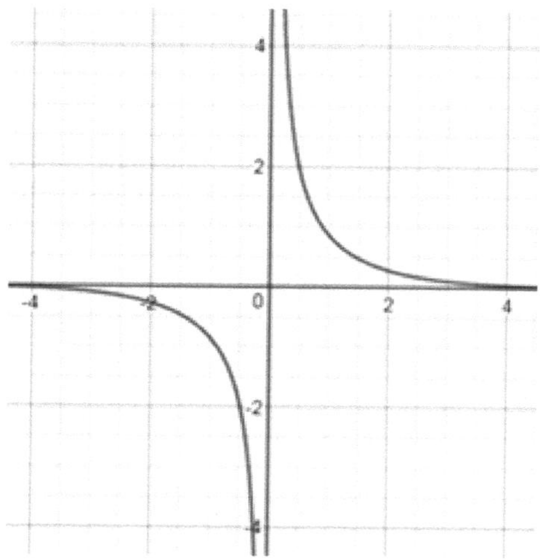

The graph of csch(x)

Note that both the x- and y-axes are asymptotes. The graph is symmetric with respect to the origin. Both the domain and range of the function are the set of real numbers excluding zero.

Logarithmic differentiation

The technique is often performed in cases where it is easier to differentiate the logarithm of a function rather than the function itself. This usually occurs in cases where the function of interest is composed of a product of a number of parts, so that a logarithmic transformation will turn it into a sum of separate parts (which is much easier to differentiate). It can also be useful when applied to functions raised to the power of variables or functions. Logarithmic differentiation relies on the chain rule as well as properties of logarithms (in particular, the natural logarithm, or the logarithm to the base e) to transform products into sums and divisions into subtractions. The principle can be implemented, at least in part, in the differentiation of almost all differentiable functions, providing that these functions are non-zero.

Implicit Functions

We're used to seeing algebraic definitions of functions in which the output y is given explicitly in terms of some combination of the input x :

$$y = \sqrt{x^2 + 1}$$

$$y = 2\cos(\pi x) - \cos(2\pi x)$$

$$y = \frac{1}{1 + 2e^{-x}}$$

We think of these rules as accepting the input x, going through a series of steps that transform x, and then announcing the final result as y. The output y is not involved in the "business" of computing the function, it is only the end product.

What, then, do we make of equations like this:

$$x^2 + y^2 = 1$$

$$y^3 + x^3 = 3xy$$

$$2^{xy} = x + y$$

Here the xs and ys are all mixed-up together, on either side of the equation. There is no obvious way to proceed from an "input" x to an "output" y. It isn't even clear if each y is actually a function of the corresponding x.

We could, of course, attempt to solve each of these equations for y, writing y explicitly as a function of x, and so try to put ourselves back on familiar ground. In all three cases, however, we will fail.

Sometimes the reason we will fail is that y is not uniquely determined by the value of x: many ys will balance the equation when we plug in a particular x. (For example, solving the first equation for y leads to $y = \pm\sqrt{1 - x^2}$.) Sometimes the reason is a failure of algebra itself: we can not solve for y; it is just too mixed-up with the xs for us to be able to extract it. (Try, for example, to solve the last equation for y.)

The ys in such "mixed-up" equations are often called implicit functions of x. Using the word "function" here is tricky, because, as we have seen, there may in fact be many ys corresponding to a particular x. What is "implicit" (the dictionary defines this as "implied or understood although not directly expressed") is the understanding that many functions may be lurking in such equations, if only we could coax them out.

One way to get at the implicit functions embedded in a "mixed-up" equation is to graph all of the pairs (x, y) that balance the equation.

For example, we have noted that if

$$x^2 + y^2 = 1$$

then

$y = f(x) = +\sqrt{1 - x^2}$ or $y = g(x) = -\sqrt{1 - x^2}$

Here the equation implicitly defines two perfectly normal functions, and it is a simple matter to plot them together:
Here the equation implicitly defines two perfectly normal functions, and it is a simple matter to plot them together:

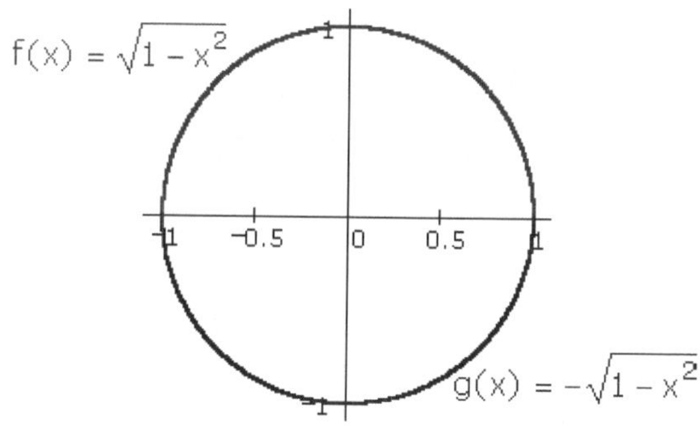

Other cases are harder. For example,

$$y^3 + x^3 = 3xy$$

This is the famous Folium of Descartes. While there is an algebraic method for solving this equation for the various functions that give y in terms of x, the results are extremely complicated.

To produce a graph of such an equation, we can divide a region of the xy-plane into a very fine grid. At each point (x, y) on the grid, we can plug the coordinate values into both the left and the right sides of the equation. If the two sides agree (within some pre-set tolerance), we add the point to the plot. If the two sides do not agree, we don't. This method may seem tedious and time consuming, but it is the kind of thing for which computers are tailor-made.

Here's a plot of the Folium of Descartes, produced in this manner:

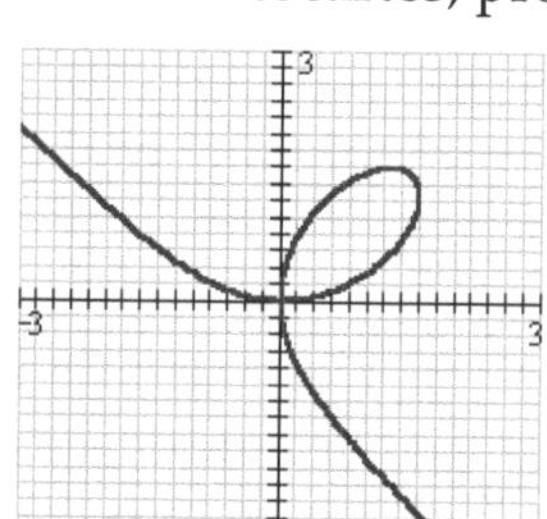

We see that the equation divides the xy-plane into three regions. Along the "tails" y can be given as an explicit function of x, but in the x

neighborhood of the "loop" three separate functions of x are implicitly defined:

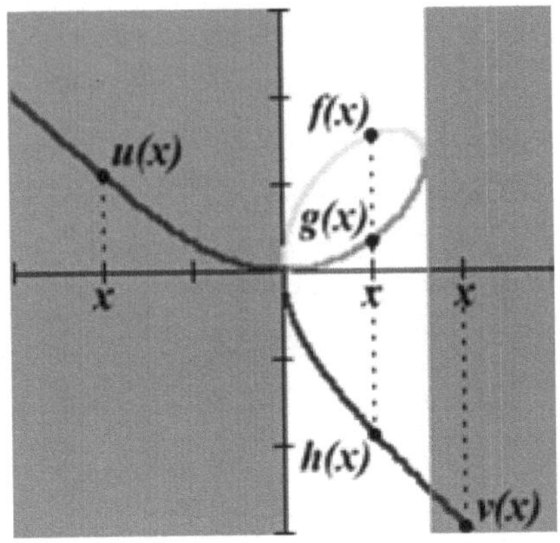

Parametric Equations And Curves

To this point (in both Calculus I and Calculus II) we've looked almost exclusively at functions in the form y=f(x)y=f(x) or x=h(y)x=h(y) and almost all of the formulas that we've developed require that functions be in one of these two forms. The problem is that not all curves or equations that we'd like to look at fall easily into this form.

Take, for example, a circle. It is easy enough to write down the equation of a circle centered at the origin with radius rr.

x2+y2=r2x2+y2=r2

However, we will never be able to write the equation of a circle down as a single equation in either of the forms above. Sure we can solve for xx or yy as the following two formulas show

y=±"r2"x2x=±"r2"y2y=±r2"x2x=±r2"y2

but there are in fact two functions in each of these. Each formula gives a portion of the circle.

y="r2"x2(top) x="r2"y2(right side)y=""r2"x2(bottom)x=""r2"y2(left side)y=r2"x2(top) x=r2"y2(right side)y="r2"x2(bottom)x="r2"y2(left side)

Unfortunately, we usually are working on the whole circle, or simply can't say that we're going to be working only on one portion of it. Even if we can narrow things down to only one of these portions the function is still often fairly unpleasant to work with.

There are also a great many curves out there that we can't even write down as a single equation in terms of only xx and yy. So, to deal with some of these problems we introduce parametric equations. Instead of defining yy in terms of xx (y=f(x)y=f(x)) or xx in terms of yy (x=h(y)x=h(y)) we define both xx and yy in terms of a third variable called a parameter as follows, x=f(t)y=g(t)x=f(t)y=g(t)

This third variable is usually denoted by tt (as we did here) but doesn't have to be of course. Sometimes we will restrict the values of tt that we'll use and at other times we won't. This will often be dependent on the problem and just what we are attempting to do.

Each value of tt defines a point (x,y)=(f(t),g(t))(x,y)=(f(t),g(t)) that we can plot. The collection of points that we get by letting tt be all possible values is the graph of the parametric equations and is called the parametric curve.

To help visualize just what a parametric curve is pretend that we have a big tank of water that is in constant motion and we drop a ping pong ball into the tank. The point (x,y)=(f(t),g(t))(x,y)=(f(t),g(t)) will then represent the location of the ping pong ball in the tank at time tt and the parametric curve will be a trace of all the locations of the ping pong ball. Note that this is not always a correct analogy but it is useful initially to help visualize just what a parametric curve is.

Sketching a parametric curve is not always an easy thing to do. Let's take a look at an example to see one way of sketching a parametric curve. This example will also illustrate why this method is usually not the best.

We should give a small warning at this point. Because of the ideas involved in them we concentrated on parametric curves that retraced portions of the curve more than once. Do not, however, get too locked into the idea that this will always happen. Many, if not most parametric curves will only trace out once. The first one we looked at is a good example of this. That parametric curve will never repeat any portion of itself.

There is one final topic to be discussed in this section before moving on. So far we've started with parametric equations and eliminated the parameter to determine the parametric curve.

However, there are times in which we want to go the other way. Given a function or equation we might want to write down a set of parametric equations for it. In these cases we say that we parameterize the function.

If we take Examples 4 and 5 as examples we can do this for ellipses (and hence circles). Given the ellipse

x2a2+y2b2=1x2a2+y2b2=1

a set of parametric equations for it would be,

x=acosty=bsintx=acosa ty=bsina t

This set of parametric equations will trace out the ellipse starting at the point (a,0)(a,0) and will trace in a counter-clockwise direction and will trace out exactly once in the range 0d"td"2À0d"td"2ð. This is a fairly important set of parametric equations as it used continually in some subjects with dealing with ellipses and/or circles. Every curve can be parameterized in more than one way. Any of the following will also parameterize the same ellipse.

x=acos(Ét)y=bsin(Ét)x=asin(Ét)y=bcos(Ét)x=acos(Ét)y="bsin(Ét)x=acosa (ùt)y=bsina (ùt)x=asina (ùt)y=bcosa (ùt)x=acosa (ùt)y="bsina (ùt)

The presence of the Éù will change the speed that the ellipse rotates as we saw in Example 5. Note as well that the last two will trace out ellipses with a clockwise direction of motion (you might want to verify this). Also note that they won't all start at the same place (if we think of t=0t=0 as the starting point that is).

There are many more parameterizations of an ellipse of course, but you get the idea. It is important to remember that each parameterization will trace out the curve once with a potentially different range of tt's. Each parameterization may rotate with different directions of motion and may start at different points.

You may find that you need a parameterization of an ellipse that starts at a particular place and has a particular direction of motion and so you now know that with some work you can write down a set of parametric equations that will give you the behavior that you're after.

Now, let's write down a couple of other important parameterizations and all the comments about direction of motion, starting point, and range of tt's for one trace (if applicable) are still true.

First, because a circle is nothing more than a special case of an ellipse we can use the parameterization of an ellipse to get the parametric equations for a circle centered at the origin of radius rr as well. One possible way to parameterize a circle is,

x=rcosty=rsintx=rcosa ty=rsina t

Finally, even though there may not seem to be any reason to, we can also parameterize functions in the form y=f(x)y=f(x) or x=h(y)x=h(y). In these cases we parameterize them in the following way,

x=tx=h(t)y=f(t)y=tx=tx=h(t)y=f(t)y=t

At this point it may not seem all that useful to do a parameterization of a function like this, but there are many instances where it will actually be easier, or it may even be required, to work with the parameterization instead of the function itself. Unfortunately, almost all of these instances occur in a Calculus III course.

6

Set Theory

Set theory is a branch of mathematical logic that studies sets, which informally are collections of objects. Although any type of object can be collected into a set, set theory is applied most often to objects that are relevant to mathematics. The language of set theory can be used to define nearly all mathematical objects.

The modern study of set theory was initiated by Georg Cantor and Richard Dedekind in the 1870s. After the discovery of paradoxes in naive set theory, such as Russell's paradox, numerous axiom systems were proposed in the early twentieth century, of which the Zermelo–Fraenkel axioms, with or without the axiom of choice, are the best-known.

Set theory is commonly employed as a foundational system for mathematics, particularly in the form of Zermelo–Fraenkel set theory with the axiom of choice. Beyond its foundational role, set theory is a branch of mathematics in its own right, with an active research community. Contemporary research into set theory includes a diverse collection of topics, ranging from the structure of the real number line to the study of the consistency of large cardinals.

HISTORY OF SET THEORY

Mathematical topics typically emerge and evolve through interactions among many researchers. Set theory, however, was founded by a single paper in 1874 by Georg Cantor: "On a Property of the Collection of All Real Algebraic Numbers".

Since the 5th century BC, beginning with Greek mathematician Zeno of Elea in the West and early Indian mathematicians in the East, mathematicians had struggled with the concept of infinity. Especially notable is the work of Bernard Bolzano in the first half of the 19th century. Modern understanding of infinity began in 1870–1874 and was motivated by Cantor's work in real analysis. An 1872 meeting between Cantor and Richard Dedekind influenced Cantor's thinking and culminated in Cantor's 1874 paper.

Cantor's work initially polarized the mathematicians of his day. While Karl Weierstrass and Dedekind supported Cantor, Leopold Kronecker, now seen as a founder of mathematical constructivism, did not. Cantorian set theory eventually became widespread, due to the utility of Cantorian concepts, such as one-to-one correspondence among sets, his proof that there are more real numbers than integers, and the "infinity of infinities" ("Cantor's paradise") resulting from the power set operation. This utility of set theory led to the article "Mengenlehre" contributed in 1898 by Arthur Schoenflies to Klein's encyclopedia.

The next wave of excitement in set theory came around 1900, when it was discovered that some interpretations of Cantorian set theory gave rise to several contradictions, called antinomies or paradoxes. Bertrand Russell and Ernst Zermelo independently found the simplest and best known paradox, now called Russell's paradox: consider "the set of all sets that are not members of themselves", which leads to a contradiction since it must be a member of itself and not a member of itself. In 1899 Cantor had himself posed the question "What is the cardinal number of the set of all sets?", and obtained a related paradox. Russell used his paradox as a theme in his 1903 review of continental mathematics in his The Principles of Mathematics.

In 1906 English readers gained the book Theory of Sets of Points by husband and wife William Henry Young and Grace Chisholm Young, published by Cambridge University Press.

The momentum of set theory was such that debate on the paradoxes did not lead to its abandonment. The work of Zermelo in 1908 and the work of Abraham Fraenkel and Thoralf Skolem in 1922 resulted in the set of axioms ZFC, which became the most commonly used set of axioms for set theory. The work of analysts such as Henri Lebesgue demonstrated the great mathematical utility of set theory, which has since become woven into the fabric of modern mathematics. Set theory is commonly used as a foundational system, although in some areas—such as algebraic geometry and algebraic topology—category theory is thought to be a preferred foundation.

Basic concepts and notation

Set theory begins with a fundamental binary relation between an object o and a set A. If o is a member (or element) of A, the notation o"A is used. A set is described by listing elements separated by commas, or by a characterizing property of its elements, within braces { }. Since sets are objects, the membership relation can relate sets as well.

A derived binary relation between two sets is the subset relation, also called set inclusion. If all the members of set A are also members of set B, then A is a subset of B, denoted A†"B. For example, {1, 2} is a subset of {1, 2, 3} , and so is {2} but {1, 4} is not. As insinuated from this definition, a set is a subset of itself. For cases where this possibility is unsuitable or would make sense to be rejected, the term proper subset is defined. A is called a proper subset of B if and only if A is a subset of B, but A is not equal to B. Note also that 1, 2, and 3 are members (elements) of the set {1, 2, 3} but are not subsets of it; and in turn, the subsets, such as {1}, are not members of the set {1, 2, 3}.

Just as arithmetic features binary operations on numbers, set theory features binary operations on sets. The:

- Union of the sets A and B, denoted A*"B, is the set of all objects that are a member of A, or B, or both. The union of {1, 2, 3} and {2, 3, 4} is the set {1, 2, 3, 4} .
- Intersection of the sets A and B, denoted A)" B, is the set of all objects that are members of both A and B. The intersection of {1, 2, 3} and {2, 3, 4} is the set {2, 3} .
- Set difference of U and A, denoted U \ A, is the set of all members of U that are not members of A. The set difference {1, 2, 3} \ {2, 3, 4} is {1} , while, conversely, the set difference {2, 3, 4} \ {1, 2, 3} is {4} . When A is a subset of U, the set difference U \ A is also called the complement of A in U. In this case, if the choice of U is clear from the context, the notation A^c is sometimes used instead of U \ A, particularly if U is a universal set as in the study of Venn diagrams.
- Symmetric difference of sets A and B, denoted A³%B or A–"B, is the set of all objects that are a member of exactly one of A and B (elements which are in one of the sets, but not in both). For instance, for the sets {1, 2, 3} and {2, 3, 4} , the symmetric difference set is {1, 4} . It is the set difference of the union and the intersection, (A*"B) \ (A)" B) or (A \ B) *" (B \ A).
- Cartesian product of A and B, denoted A × B, is the set whose members are all possible ordered pairs (a, b) where a is a member of A and b is a member of B. The cartesian product of {1, 2} and {red, white} is {(1, red), (1, white), (2, red), (2, white)}.
- Power set of a set A is the set whose members are all of the possible subsets of A. For example, the power set of {1, 2} is { {}, {1}, {2}, {1, 2} } .

Some basic sets of central importance are the empty set (the unique set containing no elements; occasionally called the null set though this name is ambiguous), the set of natural numbers, and the set of real numbers.

ALGEBRA OF SETS

The algebra of sets defines the properties and laws of sets, the set-theoretic operations of union, intersection, and complementation and the relations of set equality and set inclusion. It also provides systematic procedures for evaluating expressions, and performing calculations, involving these operations and relations.

Any set of sets closed under the set-theoretic operations forms a Boolean algebra with the join operator being union, the meet operator being intersection, the complement operator being set complement, the bottom being " and the top being the universe set under consideration.

Fundamentals

The algebra of sets is the set-theoretic analogue of the algebra of numbers. Just as arithmetic addition and multiplication are associative and commutative, so are set union and intersection; just as the arithmetic relation "less than or equal" is reflexive, antisymmetric and transitive, so is the set relation of "subset".

It is the algebra of the set-theoretic operations of union, intersection and complementation, and the relations of equality and inclusion. For a basic introduction to sets see the article on sets, for a fuller account see naive set theory, and for a full rigorous axiomatic treatment see axiomatic set theory.

The fundamental laws of set algebra

The binary operations of set union and intersection satisfy many identities. Several of these identities or "laws" have well established names.

Commutative laws:

- A *" B = B *" A Associative laws:
- (A *" B) *" C = A *" (B *" C) Distributive laws:
- A *" (B)" C) = (A *" B))" (A *" C) The union and intersection of sets may be seen as analogous to the addition and multiplication of numbers. Like addition and multiplication, the operations of union and intersection are commutative and associative, and intersection distributes over union. However, unlike addition and multiplication, union also distributes over intersection.

Two additional pairs of laws involve the special sets called the empty set Ø and the universe set U ; together with the complement operator (A C denotes the complement of A . This can also be written as A 2 , read as A prime). The empty set has no members, and the universe set has all possible members (in a particular context).

Identity laws:

- A *" " = A Complement laws:
- A *" A C = U The identity laws (together with the commutative laws) say that, just like 0 and 1 for addition and multiplication, Ø and U are the identity elements for union and intersection, respectively.

Unlike addition and multiplication, union and intersection do not have inverse elements. However the complement laws give the fundamental properties of the somewhat inverse-like unary operation of set complementation.

The preceding five pairs of laws—the commutative, associative, distributive, identity and complement laws—encompass all of set algebra, in the sense that every valid proposition in the algebra of sets can be derived from them.

Note that if the complement laws are weakened to the rule (A C) C = A, then this is exactly the algebra of propositional linear logic.

The principle of duality

Each of the identities stated above is one of a pair of identities such that each can be transformed into the other by interchanging *" and)", and also Ø and U.

These are examples of an extremely important and powerful property of set algebra, namely, the principle of duality for sets, which asserts that for any true statement about sets, the dual statement obtained by interchanging unions and intersections, interchanging U and Ø and reversing inclusions is also true. A statement is said to be self-dual if it is equal to its own dual.

Some additional laws for unions and intersections

The following proposition states six more important laws of set algebra, involving unions and intersections.

PROPOSITION 3: For any subsets A and B of a universe set U, the following identities hold:

idempotent laws:

- A *" A = A domination laws:
- A *" U = U absorption laws:
- A *" (A)" B) = A As noted above, each of the laws stated in proposition 3 can be derived from the five fundamental pairs of laws stated above. As an illustration, a proof is given below for the idempotent law for union.

Proof:

		by the identity law of intersection
		by the complement law for union
		by the distributive law of union over intersection
		by the complement law for intersection
		by the identity law for union

The following proof illustrates that the dual of the above proof is the proof of the dual of the idempotent law for union, namely the idempotent law for intersection.

Proof:

		by the identity law for union
		by the complement law for intersection
		by the distributive law of intersection over union
		by the complement law for union
		by the identity law for intersection

Intersection can be expressed in terms of set difference :

A)″ B = A ″ (A ″ B) {\displaystyle A\cap B=A\setminus (A\setminus B)}

Some additional laws for complements

The following proposition states five more important laws of set algebra, involving complements.

PROPOSITION 4: Let A and B be subsets of a universe U, then:

De Morgan's laws:

- (A *″ B) C = A C)″ B C {\displaystyle (A\cup B)^{C}=A^{C}\cap B^{C}} double complement or involution law:
- complement laws for the universe set and the empty set:
- ″ C = U {\displaystyle \varnothing ^{C}=U} Notice that the double complement law is self-dual.

The next proposition, which is also self-dual, says that the complement of a set is the only set that satisfies the complement laws. In other words, complementation is characterized by the complement laws.

PROPOSITION 5: Let A and B be subsets of a universe U, then:

uniqueness of complements:

- If A *″ B = U, and A)″ B = ″, then B = A C

The algebra of inclusion

The following proposition says that inclusion, that is the binary relation of one set being a subset of another, is a partial order.

PROPOSITION 6: If A, B and C are sets then the following hold:

reflexivity:

- A †" A antisymmetry:
- A †" B and B †" A if and only if A = B
  transitivity:
- If A †" B and B †" C , then A †" C

The following proposition says that for any set S, the power set of S, ordered by inclusion, is a bounded lattice, and hence together with the distributive and complement laws above, show that it is a Boolean algebra.

PROPOSITION 7: If A, B and C are subsets of a set S then the following hold:

existence of a least element and a greatest element:

- " †" A †" S {\displaystyle \varnothing \subseteq A\subseteq S}
  existence of joins:
- A †" A *" B If A †" C and B †" C , then A *" B †" C
  existence of meets:
- A)" B †" A If C †" A and C †" B , then C †" A)" B

The following proposition says that the statement A †" B {\displaystyle A\subseteq B} is equivalent to various other statements involving unions, intersections and complements.

PROPOSITION 8: For any two sets A and B, the following are equivalent:

- A †" B {\displaystyle A\subseteq B} The above proposition shows that the relation of set inclusion can be characterized by either of the operations of set union or set intersection, which means that the notion of set inclusion is axiomatically superfluous.

The algebra of relative complements

The following proposition lists several identities concerning relative complements and set-theoretic differences.

PROPOSITION 9: For any universe U and subsets A, B, and C of U, the following identities hold:

Duality (order theory)

In the mathematical area of order theory, every partially ordered set P gives rise to a dual (or opposite) partially ordered set which is often denoted by P^{op} or P^{d}. This dual order P^{op} is defined to be the set with the inverse order, i.e. x d" y holds in P^{op} if and only if y d" x holds in P. It is easy to see that this construction, which can be depicted by flipping the Hasse diagram for P upside down, will indeed yield a partially ordered set. In a broader sense, two posets are also said to be duals if they are dually isomorphic, i.e. if one poset is order isomorphic to the dual of the other.

The importance of this simple definition stems from the fact that every definition and theorem of order theory can readily be transferred to the dual order. Formally, this is captured by the Duality Principle for ordered sets:

If a given statement is valid for all partially ordered sets, then its dual statement, obtained by inverting the direction of all order relations and by dualizing all order theoretic definitions involved, is also valid for all partially ordered sets.

If a statement or definition is equivalent to its dual then it is said to be self-dual. Note that the consideration of dual orders is so fundamental that it often occurs implicitly when writing e″ for the dual order of d″ without giving any prior definition of this "new" symbol.

Examples

A bounded distributive lattice, and its dual

Naturally, there are a great number of examples for concepts that are dual:

- Greatest elements and least elements
- Maximal elements and minimal elements
- Least upper bounds (suprema, (″) and greatest lower bounds (infima, ‴)
- Upper sets and lower sets
- Ideals and filters
- Closure operators and kernel operators.

Examples of notions which are self-dual include:

- Being a (complete) lattice
- Monotonicity of functions
- Distributivity of lattices, i.e. the lattices for which "x,y,z: x ‴ (y (″ z) = (x ‴ y) (″ (x ‴ z) holds are exactly those for which the dual statement "x,y,z: x (″ (y ‴ z) = (x (″ y) ‴ (x (″ z) holds
- Being a Boolean algebra
- Being an order isomorphism.

Since partial orders are antisymmetric, the only ones that are self-dual are the equivalence relations.

Duality (mathematics)

In mathematics, a duality, generally speaking, translates concepts, theorems or mathematical structures into other concepts, theorems or structures, in a one-to-one fashion, often (but not always) by means of an involution operation: if the dual of A is B, then the dual of B is A. Such involutions sometimes have fixed points, so that the dual of A is A itself. For example, Desargues' theorem is self-dual in this sense under the standard duality in projective geometry. In mathematical contexts, duality has numerous meanings although it is "a very pervasive and important concept in (modern) mathematics" and "an important general theme that has manifestations in almost every area of mathematics".

Many mathematical dualities between objects of two types correspond to pairings, bilinear functions from an object of one type and another object of the second type to some family of scalars. For instance, linear algebra duality corresponds in this way to bilinear maps from pairs of vector spaces to scalars, the duality between distributions and the associated test functions corresponds to the pairing in which one integrates a distribution against a test function, and Poincaré duality corresponds similarly to intersection number, viewed as a pairing between submanifolds of a given manifold.

From a category theory viewpoint, duality can also be seen as a functor, at least in the realm of vector spaces. This functor assigns to each space its dual space, and the pullback construction assigns to each arrow f: V→W its dual f″: W″→V″.

Introductory examples

In the words of Michael Atiyah,

Duality in mathematics is not a theorem, but a "principle".

The following list of examples shows the common features of many dualities, but also indicates that the precise meaning of duality may vary from case to case.

Complement of a subset

A simple, maybe the most simple, duality arises from considering subsets of a fixed set S. To any subset A †″ S, the complement A^c consists of all those elements in S which are not contained in A. It is again a subset of S. Taking the complement has the following properties:

- Applying it twice gives back the original set, i.e., $(A^c)^c = A$. This is referred to by saying that the operation of taking the complement is an involution.
- An inclusion of sets A †″ B is turned into an inclusion in the opposite direction B^c †″ A^c.
- Given two subsets A and B of S, A is contained in B^c if and only if B is contained in A^c.

This duality appears in topology as a duality between open and closed subsets of some fixed topological space X: a subset U of X is closed if and only if its complement in X is open. Because of this, many theorems about closed sets are dual to theorems about open sets. For example, any union of open sets is open, so dually, any intersection of closed sets is closed. The interior of a set is the largest open set contained in it, and the closure of the set is the smallest closed set that contains it. Because of the duality, the complement of the interior of any set U is equal to the closure of the complement of U.

Dual cone

A set C (blue) and its dual cone C* (red).

A duality in geometry is provided by the dual cone construction. Given a set C of points in the plane R 2 (or more generally points in R n, the dual cone is defined as the set C " †" R 2 consisting of those points (x 1 , x 2) satisfying

x 1 c 1 + x 2 c 2 e" 0

for all points (c 1 , c 2) in C, as illustrated in the diagram. Unlike for the complement of sets mentioned above, it is not in general true that applying the dual cone construction twice gives back the original set C is the smallest cone containing C which may be bigger than C. Therefore this duality is weaker than the one above, in that

- Applying the operation twice gives back a possibly bigger set: for all C is contained in C " ". (For some C , namely the cones, the two are actually equal.)

  The other two properties carry over without change:
- It is still true that an inclusion C †" D is turned into an inclusion in the opposite direction (D " †" C ").
- Given two subsets C of the plane, C is contained in D " if and only if D is contained in C "

Dual vector space

A very important example of a duality arises in linear algebra by associating to any vector space V its dual vector space V*. Its elements are the k-linear maps ö : V→k {\displaystyle \varphi :V\to k} , where k is the field over which V is defined. The three properties of the dual cone carry over to this type of duality by replacing subsets of R 2 {\displaystyle \mathbb {R} ^{2}} by vector space and inclusions of such subsets by linear maps. That is:

- Applying the operation of taking the dual vector space twice gives another vector space V**. There is always a map V→V**. For some V, namely precisely the finite-dimensional vector spaces, this map is an isomorphism.
- A linear map V→W gives rise to a map in the opposite direction (W*→V*).
- Given two vector spaces V and W, the maps from V to W* correspond to the maps from W to V*.

A particular feature of this duality is that V and V^* are isomorphic for certain objects, namely finite-dimensional vector spaces. However, this is in a sense a lucky coincidence, for giving such an isomorphism requires a certain choice, for example the choice of a basis of V. This is also true in the case if V is a Hilbert space, via the Riesz representation theorem.

GALOIS THEORY

In all the dualities discussed before, the dual of an object is of the same kind as the object itself. For example, the dual of a vector space is again a vector space. Many duality statements are not of this kind. Instead, such dualities reveal a close relation between objects of seemingly different nature. One example of such a more general duality is from Galois theory. For a fixed Galois extension K / F, one may associate the Galois group Gal(K/E) to any intermediate field E (i.e., F †″ E †″ K). This group is a subgroup of the Galois group G = Gal(K/F). Conversely, to any such subgroup H †″ G there is the fixed field K^H consisting of elements fixed by the elements in H.

Compared to the above, this duality has the following features:

- An extension F †″ F2 of intermediate fields gives rise to an inclusion of Galois groups in the opposite direction: Gal(K/F2) †″ Gal(K/F).
- Associating Gal(K/E) to E and K^H to H are inverse to each other. This is the content of the fundamental theorem of Galois theory.

Order-reversing dualities

Hasse diagram of the power set of {1,2,3,4}, partially ordered by ,″. The dual poset, i.e. ordering by *f*″, is obtained by turning the diagram upside-down. The green nodes form an upper set and a lower set in the original and the dual order, respectively.

Given a poset P = (X, d″) (short for partially ordered set; i.e., a set that has a notion of ordering but in which two elements cannot necessarily be placed in order relative to each other), the dual poset P^d = (X, e″) comprises the same ground set but the converse relation. Familiar examples of dual partial orders include

- the subset and superset relations ,″ and *f*″ on any collection of sets, such as the subsets of a fixed set S. This gives rise to the first example of a duality mentioned above.
- the divides and multiple-of relations on the integers.
- the descendant-of and ancestor-of relations on the set of humans.

A duality transform is an involutive antiautomorphism f of a partially ordered set S, that is, an order-reversing involution f : S→S. In several important cases these simple properties determine the transform uniquely up to some simple symmetries. For example, if f_1, f_2 are two duality transforms then their composition is an order automorphism of S; thus, any two duality transforms differ only by an order automorphism. For example, all order automorphisms of a power set $S = 2^R$ are induced by permutations of R.

A concept defined for a partial order P will correspond to a dual concept on the dual poset P^d. For instance, a minimal element of P will be a maximal element of P^d: minimality and maximality are dual concepts in order theory. Other pairs of dual concepts are upper and lower bounds, lower sets and upper sets, and ideals and filters.

In topology, open sets and closed sets are dual concepts: the complement of an open set is closed, and vice versa. In matroid theory, the family of sets complementary to the independent sets of a given matroid themselves form another matroid, called the dual matroid.

Dimension-reversing dualities

The features of the cube and its dual octahedron correspond one-for-one with dimensions reversed.

There are many distinct but interrelated dualities in which geometric or topological objects correspond to other objects of the same type, but with a reversal of the dimensions of the features of the objects. A classical example of this is the duality of the platonic solids, in which the cube and the octahedron form a dual pair, the dodecahedron and the icosahedron form a dual pair, and the tetrahedron is self-dual. The dual polyhedron of any of these polyhedra may be formed as the convex hull of the center points of each face of the primal polyhedron, so the vertices of the dual correspond one-for-one with the faces of the primal. Similarly, each edge of the dual corresponds to an edge of the primal, and each face of the dual corresponds to a vertex of the primal. These correspondences are incidence-preserving: if two parts of the primal polyhedron touch each other, so do the corresponding two parts of the dual polyhedron. More generally, using the concept of polar reciprocation, any convex polyhedron, or more generally any convex polytope, corresponds to a dual polyhedron or dual polytope, with an i-dimensional feature of an n-dimensional polytope corresponding to an $(n - i - 1)$-dimensional feature of the dual polytope. The incidence-preserving nature of the duality is reflected in the fact that the face lattices of the primal and dual polyhedra or polytopes are themselves order-theoretic duals. Duality of polytopes and order-theoretic duality are both involutions: the dual polytope of the dual polytope of any polytope is the original polytope, and reversing all order-relations twice returns to the original order. Choosing a different center of polarity leads to geometrically different dual polytopes, but all have the same combinatorial structure.

A planar graph in blue, and its dual graph in red.

From any three-dimensional polyhedron, one can form a planar graph, the graph of its vertices and edges. The dual polyhedron has a dual graph, a graph with one vertex for each face of the polyhedron and with one edge for every two adjacent faces. The same concept of planar graph duality may be generalized to graphs that are drawn in the plane but that do not come from a three-dimensional polyhedron, or more generally to graph embeddings on surfaces of higher genus: one may draw a dual graph by placing one vertex within each region bounded by a cycle of edges in the embedding, and drawing an edge connecting any two regions that share a boundary edge. An important example of this type comes from computational geometry: the duality for any finite set S of points in the plane between the Delaunay triangulation of S and the Voronoi diagram of S. As with dual polyhedra and dual polytopes, the duality of graphs on surfaces is a dimension-reversing involution: each vertex in the primal embedded graph corresponds to a region of the dual embedding, each edge in the primal is crossed by an edge in the dual, and each region of the primal corresponds to a vertex of the dual. The dual graph depends on how the primal graph is embedded: different planar embeddings of a single graph may lead to different dual graphs. Matroid duality is an algebraic extension of planar graph duality, in the sense that the dual matroid of the graphic matroid of a planar graph is isomorphic to the graphic matroid of the dual graph.

A kind of geometric duality also occurs in optimization theory, but not one that reverses dimensions. A linear program may be specified by a system of real variables (the coordinates for a point in Euclidean space R n {\displaystyle \mathbb {R} ^{n}}), a system of linear constraints (specifying that the point lie in a halfspace; the intersection of these halfspaces is a convex polytope, the feasible region of the program), and a linear function (what to optimize). Every linear program has a dual problem with the same optimal solution, but the variables in the dual problem correspond to constraints in the primal problem and vice versa.

Duality in logic and set theory

In logic, functions or relations A and B are considered dual if A(¬x) = ¬B(x), where ¬ is logical negation. The basic duality of this type is the duality of the " and " quantifiers in classical logic. These are dual because "x.¬P(x) and ¬"x.P(x) are equivalent for all predicates P in classical logic: if there exists an x for which P fails to hold, then it is false that P holds for all x (but the converse does not hold constructively). From this fundamental logical duality follow several others:

- A formula is said to be satisfiable in a certain model if there are assignments to its free variables that render it true; it is valid if every assignment to its free variables makes it true. Satisfiability and validity are dual because the invalid formulas are precisely those whose negations are satisfiable, and the unsatisfiable formulas are those whose negations are valid. This can be viewed as a special case of the previous item, with the quantifiers ranging over interpretations.
- In classical logic, the '" and (" operators are dual in this sense, because (¬x '" ¬y) and ¬(x (" y) are equivalent. This means that for every theorem of classical logic there is an equivalent dual theorem. De Morgan's laws are examples. More generally, '" (¬ x_i) = ¬(" x_i. The left side is true if and only if "i.¬x_i, and the right side if and only if ¬"i.x_i.
- In modal logic, ¡% p means that the proposition p is "necessarily" true, and Ê% p that p is "possibly" true. Most interpretations of modal logic assign dual meanings to these two operators. For example in Kripke semantics, "p is possibly true" means "there exists some world W such that p is true in W", while "p is necessarily true" means "for all worlds W, p is true in W". The duality of ¡% and Ê% then follows from the analogous duality of " and ". Other dual modal operators behave similarly. For example, temporal logic has operators denoting "will be true at some time in the future" and "will be true at all times in the future" which are similarly dual.

Other analogous dualities follow from these:

- Set-theoretic union and intersection are dual under the set complement operator Å"^{C}. That is, A^C)" B^C = (A *" B)^C, and more generally,)" $A^C_á$ = (*" $A_á$)^C. This follows from the duality of " and ": an element x is a member of)" $A^C_á$ if and only if "á.¬x " $A_á$, and is a member of (*" $A_á$)^C if and only if ¬"á. x " $A_á$.

Dual objects

A group of dualities can be described by endowing, for any mathematical object X, the set of morphisms Hom (X, D) into some fixed object D, with a structure similar to that of X. This is sometimes called internal Hom. In general, this yields a true duality only for specific choices of D, in which case X* = Hom (X, D) is referred to as the dual of X. There is always a map from X to the bidual, that is to say, the dual of the dual,

X→X " " := (X ") " = Hom a (Hom a (X , D) , D) . {\displaystyle X\to X^{**}:=(X^{*})^{*}=\operatorname {Hom} (\operatorname {Hom} (X,D),D).}

It assigns to some x " X the map that associates to any map f : X→D (i.e., an element in Hom(X, D)) the value f(x). Depending on the concrete duality considered and also depending on the object X, this map may or may not be an isomorphism.

Dual vector spaces revisited

The construction of the dual vector space

V " = Hom a (V , K)

mentioned in the introduction is an example of such a duality. Indeed, the set of morphisms, i.e., linear maps, forms a vector space in its own right. The map $V \to V^{**}$ mentioned above is always injective. It is surjective, and therefore an isomorphism, if and only if the dimension of V is finite. This fact characterizes finite-dimensional vector spaces without referring to a basis.

Isomorphisms of V and V" and inner product spaces

A vector space V is isomorphic to V" precisely if V is finite-dimensional. In this case, such an isomorphism is equivalent to a non-degenerate bilinear form

ö : V × V→K

In this case V is called an inner product space. For example, if K is the field of real or complex numbers, any positive definite bilinear form gives rise to such an isomorphism. In Riemannian geometry, V is taken to be the tangent space of a manifold and such positive bilinear forms are called Riemannian metrics. Their purpose is to measure angles and distances. Thus, duality is a foundational basis of this branch of geometry. Another application of inner product spaces is the Hodge star which provides a correspondence between the elements of the exterior algebra. For an n-dimensional vector space, the Hodge star operator maps k-forms to (n " k)-forms. This can be used to formulate Maxwell's equations. In this guise, the duality inherent in the inner product space exchanges the role of magnetic and electric fields.

7

Tangents and Normals

A tangent to a curve is a line that touches the curve at one point and has the same slope as the curve at that point.

A normal to a curve is a line perpendicular to a tangent to the curve.

A tangent at a point on the curve is a straight line that touches the curve at that point and whose slope is equal to the gradient/derivative of the curve at that point. From the definition, you can deduce how to find the equation of the tangent to the curve at any point. Given a function $y = f(x)$, the equation of the tangent to this curve at $x = x_0$ can be found in the following way:

- Find out the gradient/derivative of the curve at the point $x = x_0$: To do this one needs to calculate dydx#x=x0. Let us call this value m, in analogy to the slope of a straight line.
- Find the equation of the straight line passing through the point $(x_0, y(x_0))$ with slope m. This is quite straightforward and can be found out as y–y1x–x1=m

You have found out the equation of the tangent to the curve at the given point!

NORMAL

A normal at a point on the curve is a straight line that intersects the curve at that point and is perpendicular to the tangent at that point. If its slope is given by n, and the slope of the tangent at that point or the value of the gradient/derivative at that point is given by m; then we have $m \times n = -1$. Steps for finding the normal to a given curve $y = f(x)$ at a point $x = x_0$:

- Find out the gradient/derivative of the curve at the point $x = x_0$: This first step is exactly the same as in the method of finding the equation of the tangent to the curve i.e. m = dydx#x=x0

- Find the slope 'n' of the normal: As the normal is perpendicular to the tangent, we have:
  n="1m
- Now, find the equation of the straight line passing through the point (x_0, $y(x_0)$) with slope n. The equation is given by:
  y–y1x–x1=n

It might be quite noticeable that both the tangents and normals to a curve go hand in hand. Both are easily derivable from one another. Now take a look at the diagram below to visualize them better, and then proceed towards the solved example to clear your doubts.

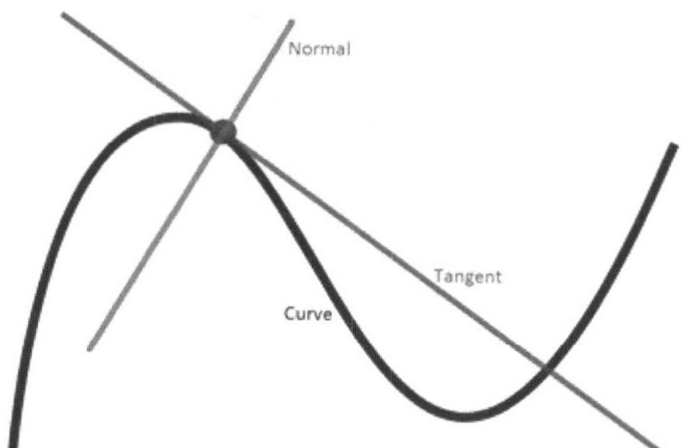

History

Euclid makes several references to the tangent. In Apollonius work Conics (c. 225 BC) he defines a tangent as being a line such that no other straight line could fall between it and the curve.

Archimedes found the tangent to an Archimedean spiral by considering the path of a point moving along the curve.

In the 1630s Fermat developed the technique of adequality to calculate tangents and other problems in analysis and used this to calculate tangents to the parabola. The technique of adeqality is similar to taking the difference between and dividing by a power of {\displaystyle h}. Independently Descartes used his method of normals based on the observation that the radius of a circle is always normal to the circle itself.

These methods led to the development of differential calculus in the 17th century. Many people contributed. Roberval discovered a general method of drawing tangents, by considering a curve as described by a moving point whose motion is the resultant of several simpler motions. René-François de Sluse and Johannes Hudde found algebraic algorithms for finding tangents. Further developments included those of John Wallis and Isaac Barrow, leading to the theory of Isaac Newton and Gottfried Leibniz.

An 1828 definition of a tangent was "a right line which touches a curve, but which when produced, does not cut it". This old definition prevents inflection points from having any tangent. It has been dismissed and the modern definitions are equivalent to those of Leibniz who defined the tangent line as the line through a pair of infinitely close points on the curve.

Tangent line to a curve

The intuitive notion that a tangent line "touches" a curve can be made more explicit by considering the sequence of straight lines (secant lines) passing through two points, A and B, those that lie on the function curve. The tangent at A is the limit when point B approximates or tends to A. The existence and uniqueness of the tangent line depends on a certain type of mathematical smoothness, known as "differentiability." For example, if two circular arcs meet at a sharp point (a vertex) then there is no uniquely defined tangent at the vertex because the limit of the progression of secant lines depends on the direction in which "point B" approaches the vertex.

At most points, the tangent touches the curve without crossing it (though it may, when continued, cross the curve at other places away from the point of tangent). A point where the tangent (at this point) crosses the curve is called an inflection point. Circles, parabolas, hyperbolasand ellipses do not have any inflection point, but more complicated curves do have, like the graph of a cubic function, which has exactly one inflection point, or a sinusoid, which has two inflection points per each period of the sine.

Conversely, it may happen that the curve lies entirely on one side of a straight line passing through a point on it, and yet this straight line is not a tangent line. This is the case, for example, for a line passing through the vertex of a triangle and not intersecting it otherwise—where the tangent line does not exist for the reasons explained above. In convex geometry, such lines are called supporting lines.

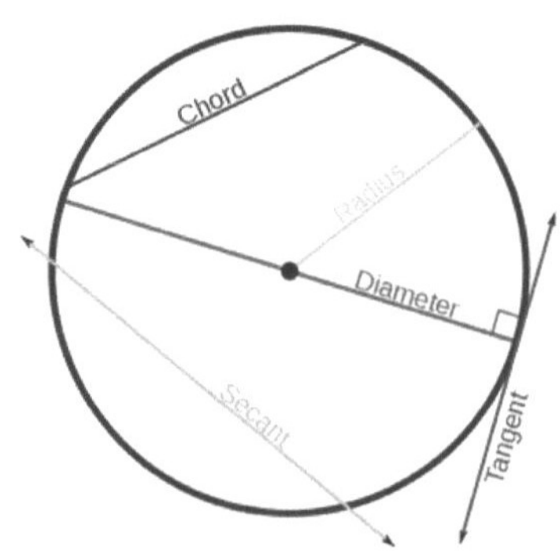

A tangent, a chord, and a secant to a circle

Analytical approach

The geometrical idea of the tangent line as the limit of secant lines serves as the motivation for analytical methods that are used to find tangent lines explicitly. The question of finding the tangent line to a graph, or the tangent line problem, was one of the central questions leading to the development of calculus in the 17th century. In the second book of his Geometry, René Descartes said of the problem of constructing the tangent to a curve, "And I dare say that this is not only the most useful and most general problem in geometry that I know, but even that I have ever desired to know".

INTUITIVE DESCRIPTION

More rigorous description

To make the preceding reasoning rigorous, one has to explain what is meant by the difference quotient approaching a certain limiting value k. The precise mathematical formulation was given by Cauchy in the 19th century and is based on the notion of limit. Suppose that the graph does not have a break or a sharp edge at pand it is neither plumb nor too wiggly near p. Then there is a unique value of k such that, as h approaches 0, the difference quotient gets closer and closer to k, and the distance between them becomes negligible compared with the size of h, if h is small enough. This leads to the definition of the slope of the tangent line to the graph as the limit of the difference quotients for the function f. This limit is the derivative of the function f at x = a, denoted f 2 (a).

How the method can fail

Calculus also demonstrates that there are functions and points on their graphs for which the limit determining the slope of the tangent line does not exist. For these points the function f is non-differentiable. There are two possible reasons for the method of finding the tangents based on the limits and derivatives to fail: either the geometric tangent exists, but it is a vertical line, which cannot be given in the point-slope form since it does not have a slope, or the graph exhibits one of three behaviors that precludes a geometric tangent.

The graph $y = x^{1/3}$ illustrates the first possibility: here the difference quotient at a = 0 is equal to $h^{1/3}/h = h^{''2/3}$, which becomes very large as h approaches 0. This curve has a tangent line at the origin that is vertical.

The graph $y = x^{2/3}$ illustrates another possibility: this graph has a cusp at the origin. This means that, when h approaches 0, the difference quotient at a = 0 approaches plus or minus infinity depending on the sign of x. Thus both branches of the curve are near to the half vertical line for which y=0, but none is near to the negative part of this line. Basically, there is no tangent at the origin in this case, but in some context one may consider this line as a tangent, and even, in algebraic geometry, as a double tangent.

The graph y = |x| of the absolute value function consists of two straight lines with different slopes joined at the origin. As a point q approaches the origin from the right, the secant line always has slope 1. As a point q approaches the origin from the left, the secant line always has slope "1. Therefore, there is no unique tangent to the graph at the origin. Having two different (but finite) slopes is called a corner.

Finally, since differentiability implies continuity, the contrapositive states discontinuity implies non-differentiability. Any such jump or point discontinuity will have no tangent line. This includes cases where one slope approaches positive infinity while the other approaches negative infinity, leading to an infinite jump discontinuity

Angle between curves

The angle between two curves at a point where they intersect is defined as the angle between their tangent lines at that point. More specifically, two curves are said to be tangent at a point if they have the same tangent at a point, and orthogonal if their tangent lines are orthogonal.

Multiple tangents at a point

The formulas above fail when the point is a singular point. In this case there may be two or more branches of the curve that pass through the point, each branch having its own tangent line. When the point is the origin, the equations of these lines can be found for algebraic curves by factoring the equation formed by eliminating all but the lowest degree terms from the original equation. Since any point can be made the origin by a change of variables, this gives a method for finding the tangent lines at any singular point.

When the curve is not self-crossing, the tangent at a reference point may still not be uniquely defined because the curve is not differentiable at that point although it is differentiable elsewhere. In this case the left and right derivatives are defined as the limits of the derivative as the point at which it is evaluated approaches the reference point from respectively the left (lower values) or the right (higher values). For example, the curve y = |x| is not differentiable at x = 0: its left and right derivatives have respective slopes –1 and 1; the tangents at that point with those slopes are called the left and right tangents.

Sometimes the slopes of the left and right tangent lines are equal, so the tangent lines coincide. This is true, for example, for the curve $y = x^{2/3}$, for which both the left and right derivatives at $x = 0$ are infinite; both the left and right tangent lines have equation $x = 0$.

Tangent to a Circle

The line that joins two infinitely close points from a point on the circle is a Tangent. In other words, we can say that the lines that intersect the circles exactly in one single point are Tangents. Point of tangency is the point where the tangent touches the circle. At the point of tangency, a tangent is perpendicular to the radius. Several theorems are related to this because it plays a significant role in geometrical constructions and proofs. We will look at them one by one.

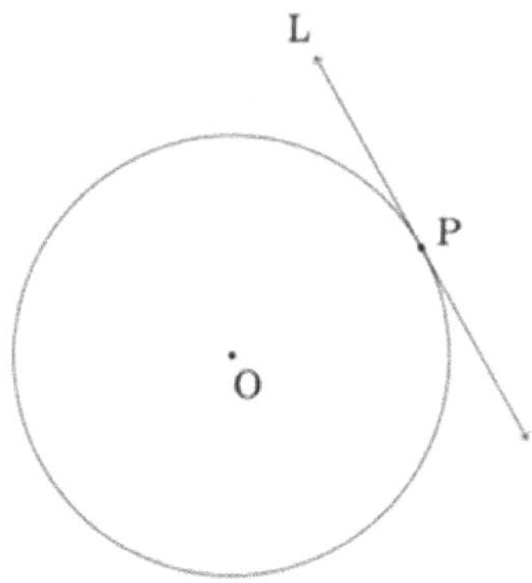

Tangents Formula

The formula for the tangent is given below. For the description of formula, please look at the following diagram.

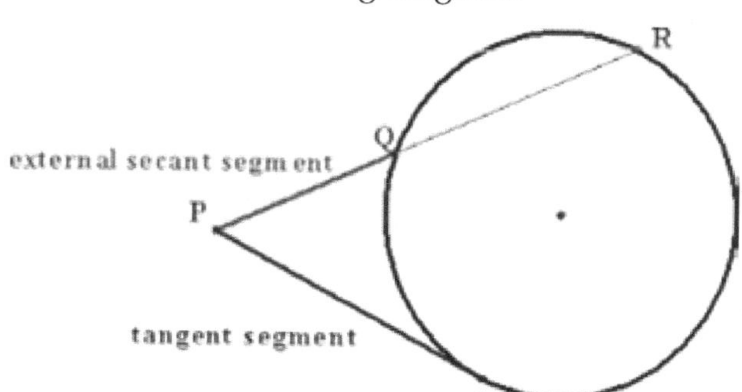

Here, we consider a circle where P is the exterior point. From that exterior point, the circle has the tangent at a points A and B. A straight line which cuts curve into two or more parts is known as a secant. So, here secant is PR is drawn and at Q, R intersects the circle as shown in the upper diagram. The formula for tangent-secant states that:

PR/PS = PS/PQ

$PS^2 = PQ.PR$

Properties of Tangents

Remember the following points about the properties of tangents-

- The tangent line never crosses the circle, it just touches the circle.
- At the point of tangency, it is perpendicular to the radius.
- A chord and tangent form an angle and this angle is same as that of tangent inscribed on the opposite side of the chord.
- From the same external point, the tangent segments to a circle are equal. Theorems for Tangents to Circle

Theorem 1

A radius is obtained by joining the centre and the point of tangency. The tangent at a point on a circle is at right angles to this radius. Just follow this below diagram: Here AB¥"OP

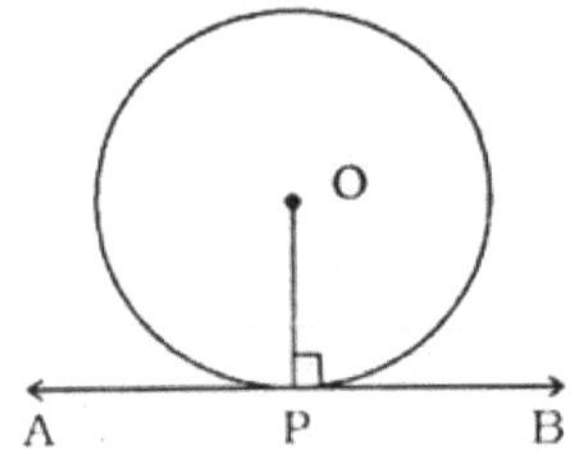

Theorem 2

This theorem states that if from one external point, two tangents are drawn to a circle then they have equal tangent segments. Tangent segment means line joining to the external point and the point of tangency. Consider the following diagram: Here, AC=BC.

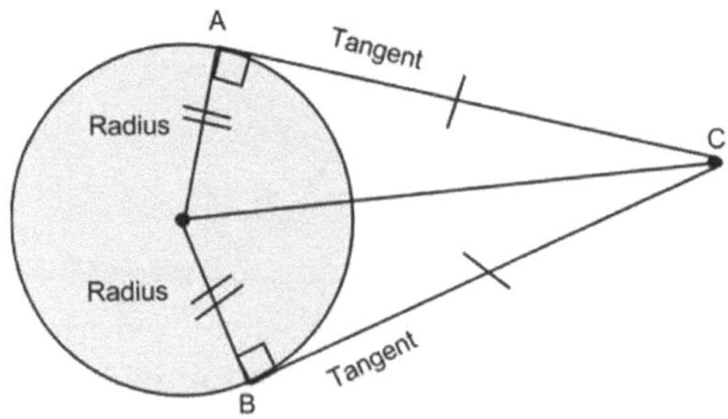

These are some of the basic theorems on tangents to a circle. In the next section, we will look at some of the basic properties of tangents.

Slope of a Straight Line

The tangent value of any trigonometric angle that a straight line makes with the positive direction of the x-axis in anticlockwise direction is called the slope or gradient of a straight line.

The angle of inclination of a line is the angle made by the line with the positive direction of the x-axis. It is usually measured from the positive x-axis in the anti-clockwise direction.

The slope of the line is generally denoted by 'm'. Thus, m = tan è. The gradient or slope of a line (not parallel to the axis of y) is the trigonometrical tangent of the angle which the line makes with the positive direction of the x-axis. Thus, if a line makes an angle è with the positive direction of the x-axis, then its slope will be tan è. The slop of a line is positive or negative according as è is acute or obtuse. Sine a line parallel to x-axis makes an angle of 0° with x-axis, therefore its slope is tan 0° = 0. A line parallel to y-axis that is i.e., perpendicular to x-axis makes an angle of 90° with x-axis, so its slope is tan À2ð2 = infinity. Also the slope of a line equally inclined with axes is 1 or -1 as it makes 45° or 135° angle with x-axis.

In short, the slope of a line is the trigonometrical tangent of its inclination.

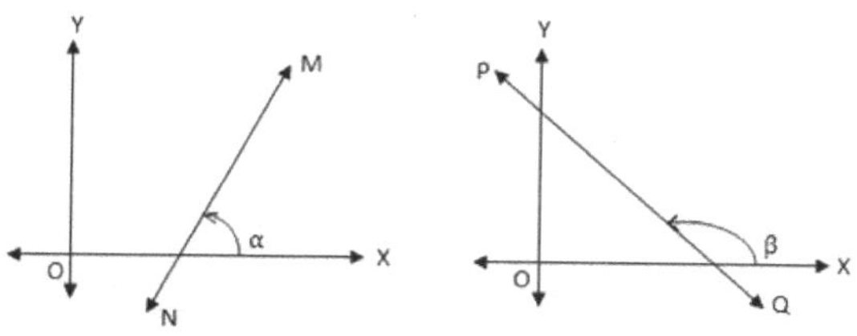

Slope of a Straight Line

In the above figure the inclination of the lines MN and PQ are á and â respectivcly.

Length of the Tangent to a Circle

Let the tangent drawn from the point $P(x_1, y_1)$ meet the circle at the point T as shown in the given diagram. The equation is given by

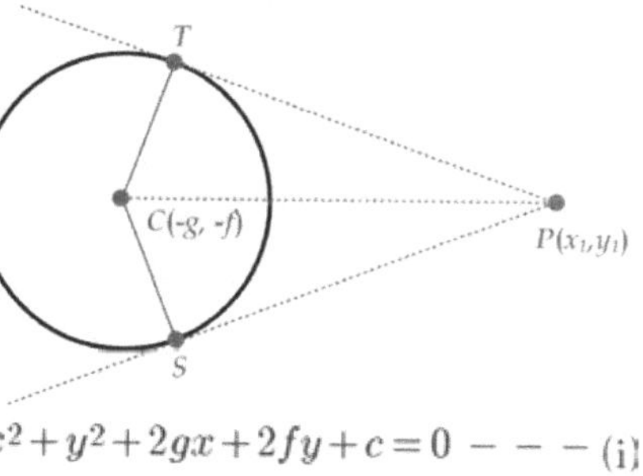

$$x^2+y^2+2gx+2fy+c=0 \quad --- (i)$$

Consider the triangle PTC formed in this way is a right triangle, so according to the given diagram we have

$$|PT|^2+|TC|^2=|PC|^2 \quad --- \text{(ii)}$$

It is observed that $|TC$ is the radius of the circle, so $|TC|^2=g^2+f^2-c.$

We also have

$$|PC|^2=(x_1-(-g))^2+(y_1-(-f))^2=(x_1+g)^2+(y_1+f)^2$$

Putting all these values in (ii), we get

This gives the length of the tangent from the point $P(x_1,y_1)$ to the circle $x^2+y^2+2gx+2fy+c=0$.

Similarly, we can show that the PS is also of the same length.

Example: Find the length of the tangent from $(12,-9)$ to the circle

$3x^2+3y^2-7x+22y+9=0$

Dividing the equation of the circle by 3, we get the standard form

$$x^2+y^2-\frac{7}{3}x+\frac{22}{3}y+3=0$$

The required length of the tangent from is

$$\sqrt{(12)^2+(-9)^2-\frac{7}{3}(12)+\frac{22}{3}(-9)+3}=\sqrt{144+81-28-66+3}=\sqrt{134}$$

Polar coordinate system

In mathematics, the polar coordinate system is a two-dimensional coordinate system in which each point on a plane is determined by a distance from a reference point and an angle from a reference direction.

The reference point (analogous to the origin of a Cartesian coordinate system) is called the pole, and the ray from the pole in the reference direction is the polar axis. The distance from the pole is called the radial coordinate or radius, and the angle is called the angular coordinate, polar angle, or azimuth.

The concepts of angle and radius were already used by ancient peoples of the first millennium BC. The Greek astronomer and astrologer Hipparchus(190–120 BC) created a table of chord functions giving the length of the chord for each angle, and there are references to his using polar coordinates in establishing stellar positions. In On Spirals, Archimedes describes the Archimedean spiral, a function whose radius depends on the angle. The Greek work, however, did not extend to a full coordinate system.

From the 8th century AD onward, astronomers developed methods for approximating and calculating the direction to Mecca (qibla)—and its distance—from any location on the Earth. From the 9th century onward they were using spherical trigonometry and map projection methods to determine these quantities accurately. The calculation is essentially the conversion of the equatorial polar coordinates of Mecca (i.e. its longitude and latitude) to its polar coordinates (i.e. its qibla and distance) relative to a system whose reference meridian is the great circle through the given location and the Earth's poles, and whose polar axis is the line through the location and its antipodal point.

There are various accounts of the introduction of polar coordinates as part of a formal coordinate system. The full history of the subject is described in Harvard professor Julian Lowell Coolidge's Origin of Polar Coordinates. Grégoire de Saint-Vincent and Bonaventura Cavalieri independently introduced the concepts in the mid-seventeenth century. Saint-Vincent wrote about them privately in 1625 and published his work in 1647, while Cavalieri published his in 1635 with a corrected version appearing in 1653. Cavalieri first used polar coordinates to solve a problem relating to the area within an Archimedean spiral. Blaise Pascal subsequently used polar coordinates to calculate the length of parabolic arcs.

In Method of Fluxions (written 1671, published 1736), Sir Isaac Newton examined the transformations between polar coordinates, which he referred to as the "Seventh Manner; For Spirals", and nine other coordinate systems. In the journal Acta Eruditorum (1691), Jacob Bernoulli used a system with a point on a line, called the pole and polar axis respectively. Coordinates were specified by the distance from the pole and the angle from the polar axis. Bernoulli's work extended to finding the radius of curvature of curves expressed in these coordinates.

The actual term polar coordinates has been attributed to Gregorio Fontana and was used by 18th-century Italian writers. The term appeared in English in George Peacock's 1816 translation of Lacroix's Differential and Integral Calculus. Alexis Clairaut was the first to think of polar coordinates in three dimensions, and Leonhard Euler was the first to actually develop them.

Conventions

The radial coordinate is often denoted by r or ñ, and the angular coordinate by ö, è, or t. The angular coordinate is specified as ö by ISO standard 31-11. However, in mathematical literature the angle is often denoted by è instead of ö.

Angles in polar notation are generally expressed in either degrees or radians (2ð rad being equal to 360°). Degrees are traditionally used in navigation, surveying, and many applied disciplines, while radians are more common in mathematics and mathematical physics.

The angle ö is defined to start at 0° from a reference direction, and to increase for rotations in either counterclockwise (ccw) or clockwise (cw) orientation. In mathematics, e.g., the reference direction is usually drawn as a ray from the pole horizontally to the right, and the polar angle increases to positive angles for ccw rotations, whereas in navigation (bearing, heading) the 0°-heading is drawn vertically upwards and the angle increases for cw rotations. The polar angles decrease towards negative values for rotations in the respectively opposite orientations.

Uniqueness of polar coordinates

Adding any number of full turns (360°) to the angular coordinate does not change the corresponding direction. Similarly, any polar coordinate is identical to the coordinate with the negative radial component and the opposite direction (adding 180° to the polar angle). Therefore, the same point (r, ö) can be expressed with an infinite number of different polar coordinates (r, ö + n × 360°) and ("r, ö + 180° + n × 360°) = ("r, ö + (2n + 1) × 180°), where n is an arbitrary integer. Moreover, the pole itself can be expressed as (0, ö) for any angle ö.

Where a unique representation is needed for any point besides the pole, it is usual to limit r to positive numbers (r > 0) and ö to the interval [0, 360°) or ("180°, 180°] (in radians, [0, 2ð) or ("ð, ð]). Another convention, in reverence to the usual codomain of the arctan-function, is to allow for arbitrary nonzero real values of the radial component and restrict the polar angle to ("90°, 90°]. In all cases a unique azimuth for the pole (r = 0) must be chosen, e.g., ö = 0.

Polar equation of a curve

The equation defining an algebraic curve expressed in polar coordinates is known as a polar equation. In many cases, such an equation can simply be specified by defining r as a function of ö. The resulting curve then consists of points of the form (r(ö), ö) and can be regarded as the graph of the polar function r. Note that, in contrast to Cartesian coordinates, the independent variable ö is the secondentry in the ordered pair.

Different forms of symmetry can be deduced from the equation of a polar function r. If r("ö) = r(ö) the curve will be symmetrical about the horizontal (0°/180°) ray, if r(ð " ö) = r(ö) it will be symmetric about the vertical (90°/270°) ray, and if r(ö " á) = r(ö) it will be rotationally symmetric by á clockwise and counterclockwise about the pole.

Because of the circular nature of the polar coordinate system, many curves can be described by a rather simple polar equation, whereas their Cartesian form is much more intricate. Among the best known of these curves are the polar rose, Archimedean spiral, lemniscate, limaçon, and cardioid.

For the circle, line, and polar rose below, it is understood that there are no restrictions on the domain and range of the curve.

Up to this point we've dealt exclusively with the Cartesian (or Rectangular, or x-y) coordinate system. However, as we will see, this is not always the easiest coordinate system to work in. So, in this section we will start looking at the polar coordinate system.

Coordinate systems are really nothing more than a way to define a point in space. For instance in the Cartesian coordinate system at point is given the coordinates (x,y)(x,y) and we use this to define the point by starting at the origin and then moving xx units horizontally followed by yy units vertically. This is shown in the sketch below.

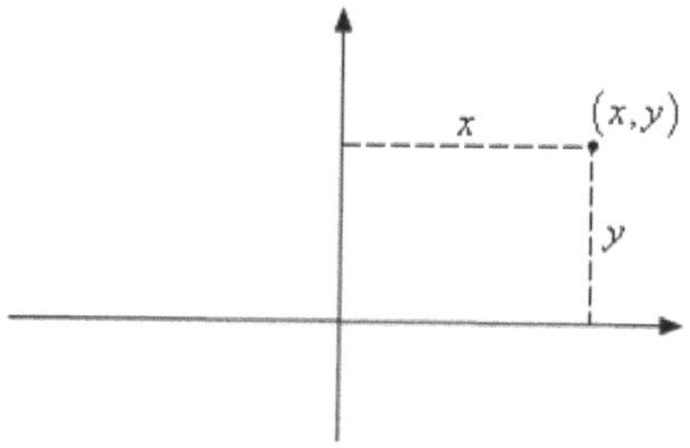

This is not, however, the only way to define a point in two dimensional space. Instead of moving vertically and horizontally from the origin to get to the point we could instead go straight out of the origin until we hit the point and then determine the angle this line makes with the positive xx-axis. We could then use the distance of the point from the origin and the amount we needed to rotate from the positive xx-axis as the coordinates of the point. This is shown in the sketch below.

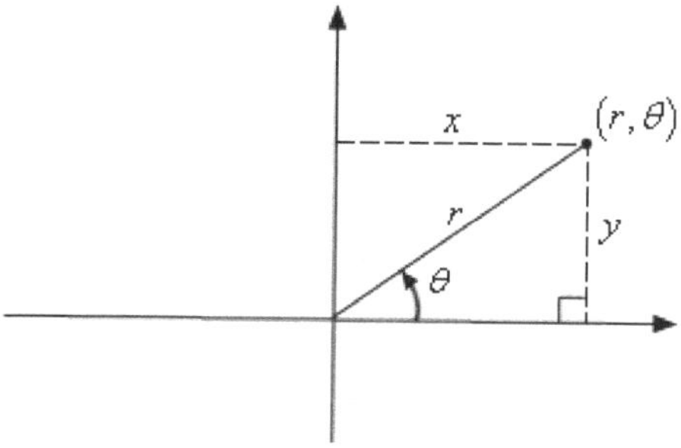

Coordinates in this form are called polar coordinates.

The above discussion may lead onc to think that rr must be a positive number. However, we also allow rr to be negative. Below is a sketch of the two points (2,À6)(2,ð6) and ("2,À6)("2,ð6).

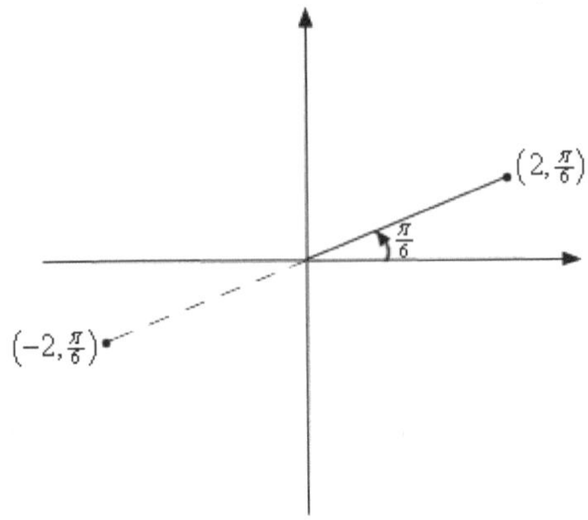

From this sketch we can see that if rr is positive the point will be in the same quadrant as ¸è. On the other hand if rr is negative the point will end up in the quadrant exactly opposite ¸è. Notice as well that the coordinates (“2,À6)(“2,ð6) describe the same point as the coordinates (2,7À6)(2,7ð6) do. The coordinates (2,7À6)(2,7ð6) tells us to rotate an angle of 7À67ð6 from the positive xx-axis, this would put us on the dashed line in the sketch above, and then move out a distance of 2.

This leads to an important difference between Cartesian coordinates and polar coordinates. In Cartesian coordinates there is exactly one set of coordinates for any given point. With polar coordinates this isn't true. In polar coordinates there is literally an infinite number of coordinates for a given point. For instance, the following four points are all coordinates for the same point.

(5,À3)=(5,″5À3)=(“5,4À3)=(“5,″2À3)(5,ð3)=(5,″5ð3)=(“5,4ð3)=(“5,″2ð3)

Here is a sketch of the angles used in these four sets of coordinates.

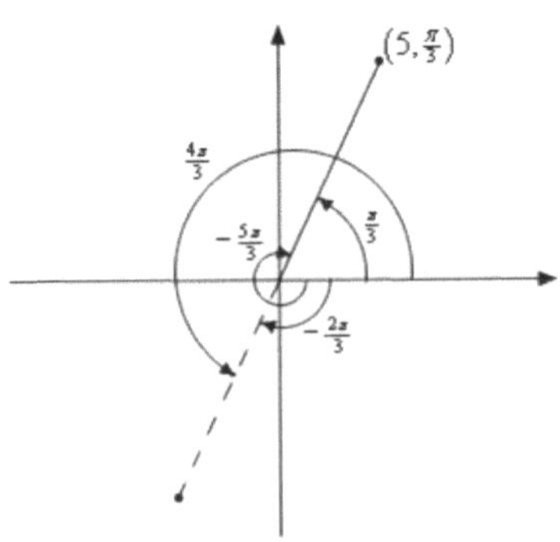

In the second coordinate pair we rotated in a clock-wise direction to get to the point. We shouldn't forget about rotating in the clock-wise direction. Sometimes it's what we have to do.

The last two coordinate pairs use the fact that if we end up in the opposite quadrant from the point we can use a negative rr to get back to the point and of course there is both a counter clock-wise and a clock-wise rotation to get to the angle.

These four points only represent the coordinates of the point without rotating around the system more than once. If we allow the angle to make as many complete rotations about the axis system as we want then there are an infinite number of coordinates for the same point. In fact, the point (r,¸)(r,è) can be represented by any of the following coordinate pairs.

Next, we should talk about the origin of the coordinate system. In polar coordinates the origin is often called the pole. Because we aren't actually moving away from the origin/pole we know that r=0r=0. However, we can still rotate around the system by any angle we want and so the coordinates of the origin/pole are (0,¸)(0,è).

Now that we've got a grasp on polar coordinates we need to think about converting between the two coordinate systems. Well start out with the following sketch reminding us how both coordinate systems work.

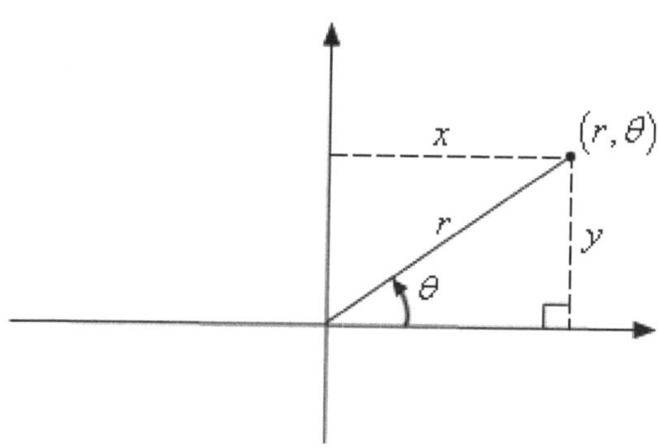

Note that we've got a right triangle above and with that we can get the following equations that will convert polar coordinates into Cartesian coordinates.

Polar to Cartesian Conversion Formulas

x=rcos¸y=rsin¸x=rcosa èy=rsina è

Converting from Cartesian is almost as easy. Let's first notice the following.

x2+y2=(rcos¸)2+(rsin¸)2=r2cos2¸+r2sin2¸=r2(cos2¸+sin2¸)=r2x2+y2=(rcosa è)2+(rsina è)2=r2cos2è+r2sin2è=r2(cos2è+sin2è)=r2

This is a very useful formula that we should remember, however we are after an equation for rr so let's take the square root of both sides. This gives, r=″x2+y2r=x2+y2

Note that technically we should have a plus or minus in front of the root since we know that rr can be either positive or negative. We will run with the convention of positive rr here.

Getting an equation for ¸è is almost as simple. We'll start with,

yx=rsin¸rcos¸=tan¸yx=rsina èrcosa è=tana è

Taking the inverse tangent of both sides gives,

¸=tan″1(yx)è=tan″1(yx)

We will need to be careful with this because inverse tangents only return values in the range ″À2<¸<À2"ð2<è<ð2. Recall that there is a second possible angle and that the second angle is given by ¸+Àè+ð.

Summarizing then gives the following formulas for converting from Cartesian coordinates to polar coordinates.

Any time we display a shape in ProcessingJS, we have to specify a pixel location, a set of x and y coordinates. These coordinates are known as Cartesian coordinates, named for René Descartes, the French mathematician who developed the ideas behind Cartesian space.

Another useful coordinate system known as polar coordinates describes a point in space as an angle of rotation around the origin and a radius from the origin. Thinking about this in terms of a vector:

- Cartesian coordinate—the x,y components of a vector
- Polar coordinate—the magnitude (length) and direction (angle) of a vector

However, the drawing functions in ProcessingJS don't understand polar coordinates. Whenever we want to display something in ProcessingJS, we have to specify locations as (x,y) Cartesian coordinates. However, sometimes it is a great deal more convenient for us to think in polar coordinates when designing. Happily for us, with trigonometry we can convert back and forth between polar and Cartesian, which allows us to design with whatever coordinate system we have in mind but always draw with Cartesian coordinates.

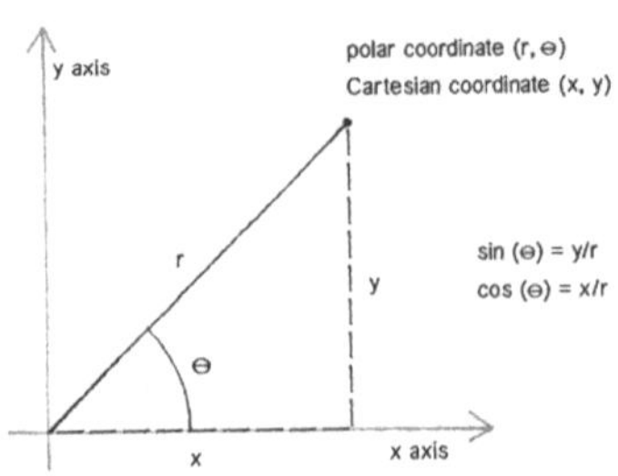

The Greek letter ¸ (theta) is often used to denote an angle, and a polar coordinate is conventionally referred to as (r, ¸) instead of (x, y). Thus, when dealing with polar coordinates, we'll now use "theta" as the preferred variable name for the angle.

sine(theta) = y/r→y = r * sine(theta)

cosine(theta) = x/r→x = r * cosine(theta)

For example, if r is 75 and theta is 45 degrees (or PI/4 radians), we can calculate x and y as below. The functions for sine and cosine in ProcessingJS are sin() and cos(), respectively. They each take one argument, an angle measured in degrees.

```
var r = 75;
var theta = 45;
// Convert polar to cartesian
var x = r * cos(theta);
var y = r * sin(theta);
```

This type of conversion can be useful in certain applications. For example, to move a shape along a circular path using Cartesian coordinates is not so easy. With polar coordinates, on the other hand, it's simple: increment the angle!

Conversion formulas

We can calculate the Cartesian coordinates of a point with polar coordinates (r,¸)(r,è) by forming the right triangle illustrated in the below figure. The hypotenuse is the line segment from the origin to the point, and its length is rr. The projection of this line segment on the xx-axis is the leg of the triangle adjacent to the angle ¸è, so x=rcos¸x=rcosa è. The yy-component is determined by the other leg, so y=rsin¸y=rsina è. Our conversion formula is

xy=rcos¸=rsin¸.(1)x=rcosa è(1)y=rsina è.

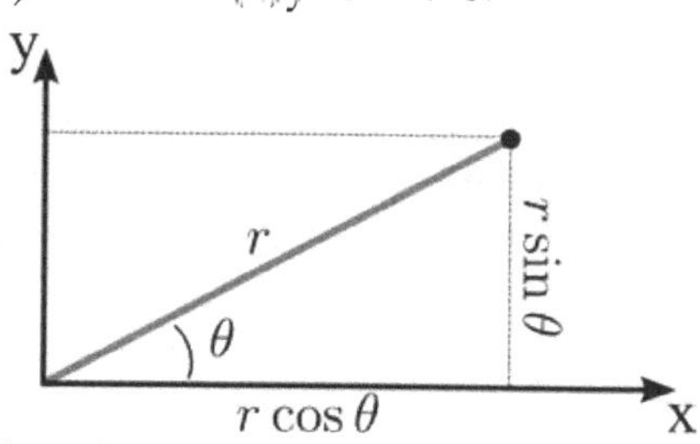

To go the other direction, one can use the same right triangle. Since rr is the distance from the origin to (x,y)(x,y), it is the magnitude r=x2+y2"""""""""r=x2+y2. Alternatively, from the equation (1)(1), one can calculate directly that

x2+y2=r2cos2¸+r2sin2¸=r2(cos2¸+sin2¸)=r2.x2+y2=r2cos2a è+r2sin2a è=r2(cos2a è+sin2a è)=r2.

Taking the ratio of yy and xx from equation (1)(1), one can obtain a formula for ¸è,

yx=rsin¸rcos¸=tan¸.yx=rsina èrcosa è=tana è.

One can also see this relationshp from the above right triangle. We can set ¸=arctanyxè=arctana yx, but a problem is that arctanarctan gives a value between ″À/2"ð/2 and À/2ð/2. One might neet to add Àð or 2À2ð to get the correct angle. With this caveat (and also mapping points where x=0x=0 to ¸=À/2è=ð/2 or ″À/2"ð/2), one obtains the following formula to convert from Cartesian to polar coordinates

r¸=x2+y2"″″″″″″″″=arctanyx.(2)r=x2+y2(2)è=arctana yx.

What about the point where (x,y)=(0,0)(x,y)=(0,0)? In this case, the angle ¸è isn′t well defined. You could just take it to be ¸=0è=0 to be concrete.

Applications

Polar coordinates are two-dimensional and thus they can be used only where point positions lie on a single two-dimensional plane. They are most appropriate in any context where the phenomenon being considered is inherently tied to direction and length from a center point. For instance, the examples above show how elementary polar equations suffice to define curves—such as the Archimedean spiral—whose equation in the Cartesian coordinate system would be much more intricate. Moreover, many physical systems—such as those concerned with bodies moving around a central point or with phenomena originating from a central point—are simpler and more intuitive to model using polar coordinates. The initial motivation for the introduction of the polar system was the study of circular and orbital motion.

Position and navigation

Polar coordinates are used often in navigation as the destination or direction of travel can be given as an angle and distance from the object being considered. For instance, aircraftuse a slightly modified version of the polar coordinates for navigation. In this system, the one generally used for any sort of navigation, the 0° ray is generally called heading 360, and the angles continue in a clockwise direction, rather than counterclockwise, as in the mathematical system. Heading 360 corresponds to magnetic north, while headings 90, 180, and 270 correspond to magnetic east, south, and west, respectively. Thus, an aircraft traveling 5 nautical miles due east will be traveling 5 units at heading 90 (read zero-niner-zero by air traffic control).

Modeling

Systems displaying radial symmetry provide natural settings for the polar coordinate system, with the central point acting as the pole. A prime example of this usage is the groundwater flow equation when applied to radially symmetric wells. Systems with a radial force are also good candidates for the use of the polar coordinate system. These systems include gravitational fields, which obey the inverse-square law, as well as systems with point sources, such as radio antennas.

Radially asymmetric systems may also be modeled with polar coordinates. For example, a microphone's pickup pattern illustrates its proportional response to an incoming sound from a given direction, and these patterns can be represented as polar curves. The curve for a standard cardioid microphone, the most common unidirectional microphone, can be represented as r = 0.5 + 0.5sin(Õ) at its target design frequency. The pattern shifts toward omnidirectionality at lower frequencies.

Angle between two curves

In order to measure the angle between two curves, we measure the angle between the tangents to the curves at that point. And that is obtained by the formula below:

$$\left|\frac{m_1 - m_2}{1 + m_1 m_2}\right|$$

where è is the angle between the 2 curves, and m_1 and m_2 are slopes or gradients of the tangents to the curve at the point of intersection.

Let's look at this through an example below:

Find the acute angle between the curves $y = x^2$ and $y = (x - 3)^2$

The curve intersects at the point (where the two equations are equal). So

$x^2 = (x - 3)^2$

$x^2 = x^2 - 6x + 9$

$6x - 9 = 0$

$$x = \frac{9}{6} = \frac{3}{2}$$

$$y = x^2 = \left(\frac{3}{2}\right)^2 = \frac{9}{4}$$

The intersection point is $\left(\frac{3}{2}, \frac{9}{4}\right)$

The tangent to $y = x^2$ at $\left(\frac{3}{2}, \frac{9}{4}\right)$ has a gradient of

$$\frac{dy}{dx} = 2x = 2\times\frac{3}{2} = 3 = m_1$$

The tangent to $y = (x-3)^2$ $\left(\frac{3}{2}, \frac{9}{4}\right)$ at has a gradient of

$$= 2(x-3)^1 = 2.\left(\frac{3}{2} - 3\right) = -2.\left(\frac{3}{2}\right) = -3 = m_2$$

$$\tan è = \left|\frac{m_1 - m_2}{1 + m_1 m_2}\right|$$

$$= \left|\frac{3 - (-3)}{1 + -3\times 3}\right|$$

$$= \left|\frac{3 + 3}{1 - 3\times 3}\right|$$

$$= \left|\frac{6}{1 - 9}\right|$$

$$= \left|\frac{-6}{8}\right|$$

$$= \frac{6}{8} = \frac{3}{4}$$

$$\tan è = \frac{6}{8}$$

$$\tan^{-1}\left(\frac{3}{4}\right)$$

$$\text{è} = \tan^{-1}\left(\frac{3}{4}\right)$$

= 36° 522 (to the nearest minute)

Differential coefficient

8

Geometry

Geometry (from the Ancient Greek: ãåùìåôñßá; geo- "earth", -metron "measurement") is a branch of mathematics concerned with questions of shape, size, relative position of figures, and the properties of space. A mathematician who works in the field of geometry is called a geometer.

Geometry arose independently in a number of early cultures as a practical way for dealing with lengths, areas, and volumes. Geometry began to see elements of formal mathematical science emerging in the West as early as the 6th century BC. By the 3rd century BC, geometry was put into an axiomatic form by Euclid, whose treatment, Euclid's Elements, set a standard for many centuries to follow.Geometry arose independently in India, with texts providing rules for geometric constructions appearing as early as the 3rd century BC. Islamic scientists preserved Greek ideas and expanded on them during the Middle Ages. By the early 17th century, geometry had been put on a solid analytic footing by mathematicians such as René Descartes and Pierre de Fermat. Since then, and into modern times, geometry has expanded into non-Euclidean geometry and manifolds, describing spaces that lie beyond the normal range of human experience.

While geometry has evolved significantly throughout the years, there are some general concepts that are more or less fundamental to geometry. These include the concepts of points, lines, planes, surfaces, angles, and curves, as well as the more advanced notions of manifolds and topology or metric.

Geometry has applications to many fields, including art, architecture, physics, as well as to other branches of mathematics.

OVERVIEW

Contemporary geometry has many subfields:

- Euclidean geometry is geometry in its classical sense. The mandatory educational curriculum of the majority of nations includes the study of points, lines, planes, angles, triangles, congruence, similarity, solid figures, circles, and analytic geometry. Euclidean geometry also has applications in computer science, crystallography, and various branches of modern mathematics.

- Differential geometry uses techniques of calculus and linear algebra to study problems in geometry. It has applications in physics, including in general relativity.
- Topology is the field concerned with the properties of geometric objects that are unchanged by continuous mappings. In practice, this often means dealing with large-scale properties of spaces, such as connectedness and compactness.
- Convex geometry investigates convex shapes in the Euclidean space and its more abstract analogues, often using techniques of real analysis. It has close connections to convex analysis, optimization and functional analysis and important applications in number theory.
- Algebraic geometry studies geometry through the use of multivariate polynomials and other algebraic techniques. It has applications in many areas, including cryptography and string theory.
- Discrete geometry is concerned mainly with questions of relative position of simple geometric objects, such as points, lines and circles. It shares many methods and principles with combinatorics.
- Computational geometry deals with algorithms and their implementations for manipulating geometrical objects. Although being a young area of geometry, it has many applications in computer vision, image processing, computer-aided design, medical imaging, etc.

History

A European and an Arab practicing geometry in the 15th century.

The earliest recorded beginnings of geometry can be traced to ancient Mesopotamia and Egypt in the 2nd millennium BC. Early geometry was a collection of empirically discovered principles concerning lengths, angles, areas, and volumes, which were developed to meet some practical need in surveying, construction, astronomy, and various crafts. The earliest known texts on geometry are the Egyptian Rhind Papyrus (2000–1800 BC) and Moscow Papyrus (c. 1890 BC), the Babylonian clay tablets such as Plimpton 322 (1900 BC). For example, the Moscow Papyrus gives a formula for calculating the volume of a truncated pyramid, or frustum. Later clay tablets (350–50 BC) demonstrate that Babylonian astronomers implemented trapezoid procedures for computing Jupiter's position and motion within time-velocity space. These geometric procedures anticipated the Oxford Calculators, including the mean speed theorem, by 14 centuries. South of Egypt the ancient Nubians established a system of geometry including early versions of sun clocks.

In the 7th century BC, the Greek mathematician Thales of Miletus used geometry to solve problems such as calculating the height of pyramids and the distance of ships from the shore. He is credited with the first use of deductive reasoning applied to geometry, by deriving four corollaries to Thales' Theorem. Pythagoras established the Pythagorean School, which is credited with the first proof of the Pythagorean theorem, though the statement of the theorem has a long history. Eudoxus (408–c. 355 BC) developed the method of exhaustion, which allowed the calculation of areas and volumes of curvilinear figures, as well as a theory of ratios that avoided the problem of incommensurable magnitudes, which enabled subsequent geometers to make significant advances. Around 300 BC, geometry was revolutionized by Euclid, whose Elements, widely considered the most successful and influential textbook of all time, introduced mathematical rigor through the axiomatic method and is the earliest example of the format still used in mathematics today, that of definition, axiom, theorem, and proof. Although most of the contents of the Elements were already known, Euclid arranged them into a single, coherent logical framework. The Elements was known to all educated people in the West until the middle of the 20th century and its contents are still taught in geometry classes today. Archimedes (c. 287–212 BC) of Syracuse used the method of exhaustion to calculate the area under the arc of a parabola with the summation of an infinite series, and gave remarkably accurate approximations of Pi. He also studied the spiral bearing his name and obtained formulas for the volumes of surfaces of revolution.

Woman teaching geometry. Illustration at the beginning of a medieval translation of Euclid's Elements, (c. 1310)

Indian mathematicians also made many important contributions in geometry. The Satapatha Brahmana (3rd century BC) contains rules for ritual geometric constructions that are similar to the Sulba Sutras. According to (Hayashi 2005, p. 363), the Úulba Sûtras contain "the earliest extant verbal expression of the Pythagorean Theorem in the world, although it had already been known to the Old Babylonians. They contain lists of Pythagorean triples, which are particular cases of Diophantine equations. In the Bakhshali manuscript, there is a handful of geometric problems (including problems about volumes of irregular solids). The Bakhshali manuscript also "employs a decimal place value system with a dot for zero." Aryabhata's Aryabhatiya (499) includes the computation of areas and volumes. Brahmagupta wrote his astronomical work Brâhma Sphuma Siddhânta in 628. Chapter 12, containing 66 Sanskrit verses, was divided into two sections: "basic operations" (including cube roots, fractions, ratio and proportion, and barter) and "practical mathematics" (including mixture, mathematical series, plane figures, stacking bricks, sawing of timber, and piling of grain). In the latter section, he stated his famous theorem on the diagonals of a cyclic quadrilateral. Chapter 12 also included a formula for the area of a cyclic quadrilateral (a generalization of Heron's formula), as well as a complete description of rational triangles (i.e. triangles with rational sides and rational areas).

In the Middle Ages, mathematics in medieval Islam contributed to the development of geometry, especially algebraic geometry. Al-Mahani (b. 853) conceived the idea of reducing geometrical problems such as duplicating the cube to problems in algebra. Thâbit ibn Qurra (known as Thebit in Latin) (836–901) dealt with arithmetic operations applied to ratios of geometrical quantities, and contributed to the development of analytic geometry. Omar Khayyám (1048–1131) found geometric solutions to cubic equations. The theorems of Ibn al-Haytham (Alhazen), Omar Khayyam and

Nasir al-Din al-Tusi on quadrilaterals, including the Lambert quadrilateral and Saccheri quadrilateral, were early results in hyperbolic geometry, and along with their alternative postulates, such as Playfair's axiom, these works had a considerable influence on the development of non-Euclidean geometry among later European geometers, including Witelo (c. 1230–c. 1314), Gersonides (1288–1344), Alfonso, John Wallis, and Giovanni Girolamo Saccheri.

In the early 17th century, there were two important developments in geometry. The first was the creation of analytic geometry, or geometry with coordinates and equations, by René Descartes (1596–1650) and Pierre de Fermat (1601–1665). This was a necessary precursor to the development of calculus and a precise quantitative science of physics. The second geometric development of this period was the systematic study of projective geometry by Girard Desargues (1591–1661). Projective geometry is a geometry without measurement or parallel lines, just the study of how points are related to each other.

Two developments in geometry in the 19th century changed the way it had been studied previously. These were the discovery of non-Euclidean geometries by Nikolai Ivanovich Lobachevsky, János Bolyai and Carl Friedrich Gauss and of the formulation of symmetry as the central consideration in the Erlangen Programme of Felix Klein (which generalized the Euclidean and non-Euclidean geometries). Two of the master geometers of the time were Bernhard Riemann (1826–1866), working primarily with tools from mathematical analysis, and introducing the Riemann surface, and Henri Poincaré, the founder of algebraic topology and the geometric theory of dynamical systems. As a consequence of these major changes in the conception of geometry, the concept of "space" became something rich and varied, and the natural background for theories as different as complex analysisand classical mechanics.

Important concepts in geometry

Axioms

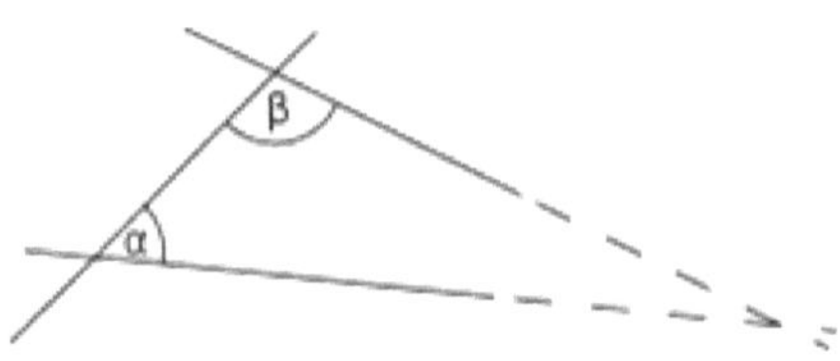

An illustration of Euclid's parallel postulate

Euclid took an abstract approach to geometry in his Elements, one of the most influential books ever written. Euclid introduced certain axioms, or postulates, expressing primary or self-evident properties of points, lines, and planes. He proceeded to rigorously deduce other properties by mathematical reasoning. The characteristic feature of Euclid's approach to geometry was its rigor, and it has come to be known as axiomatic or synthetic geometry. At the start of the 19th century, the discovery of non-Euclidean geometries by Nikolai Ivanovich Lobachevsky (1792–1856), János Bolyai (1802–1860), Carl Friedrich Gauss (1777–1855) and others led to a revival of interest in this discipline, and in the 20th century, David Hilbert (1862–1943) employed axiomatic reasoning in an attempt to provide a modern foundation of geometry.

Points

Points are considered fundamental objects in Euclidean geometry. They have been defined in a variety of ways, including Euclid's definition as 'that which has no part' and through the use of algebra or nested sets. In many areas of geometry, such as analytic geometry, differential geometry, and topology, all objects are considered to be built up from points. However, there has been some study of geometry without reference to points.

Lines

Euclid described a line as "breadthless length" which "lies equally with respect to the points on itself". In modern mathematics, given the multitude of geometries, the concept of a line is closely tied to the way the geometry is described. For instance, in analytic geometry, a line in the plane is often defined as the set of points whose coordinates satisfy a given linear equation, but in a more abstract setting, such as incidence geometry, a line may be an independent object, distinct from the set of points which lie on it. In differential geometry, a geodesic is a generalization of the notion of a line to curved spaces.

Planes

A plane is a flat, two-dimensional surface that extends infinitely far. Planes are used in every area of geometry. For instance, planes can be studied as a topological surfacewithout reference to distances or angles; it can be studied as an affine space, where collinearity and ratios can be studied but not distances; it can be studied as the complex plane using techniques of complex analysis; and so on.

Angles

Euclid defines a plane angle as the inclination to each other, in a plane, of two lines which meet each other, and do not lie straight with respect to each other. In modern terms, an angle is the figure formed by two rays, called the sides of the angle, sharing a common endpoint, called the vertex of the angle.

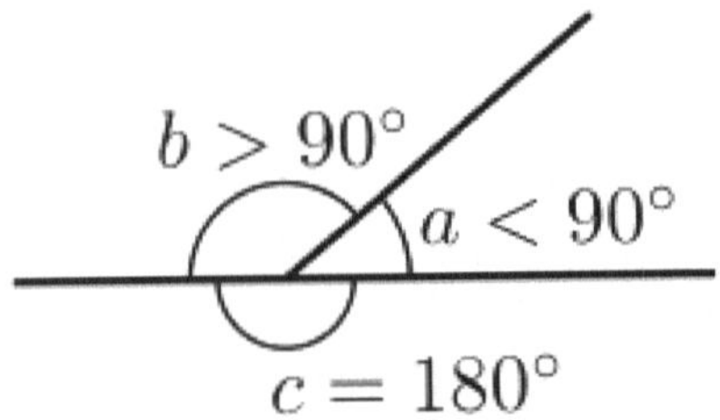

Acute (a), obtuse (b), and straight (c) angles. The acute and obtuse angles are also known as oblique angles.

In Euclidean geometry, angles are used to study polygons and triangles, as well as forming an object of study in their own right. The study of the angles of a triangle or of angles in a unit circle forms the basis of trigonometry.

In differential geometry and calculus, the angles between plane curves or space curves or surfaces can be calculated using the derivative.

Curves

A curve is a 1-dimensional object that may be straight (like a line) or not; curves in 2-dimensional space are called plane curves and those in 3-dimensional space are called space curves.

In topology, a curve is defined by a function from an interval of the real numbers to another space. In differential geometry, the same definition is used, but the defining function is required to be differentiable Algebraic geometry studies algebraic curves, which are defined as algebraic varieties of dimensionone.

Surfaces

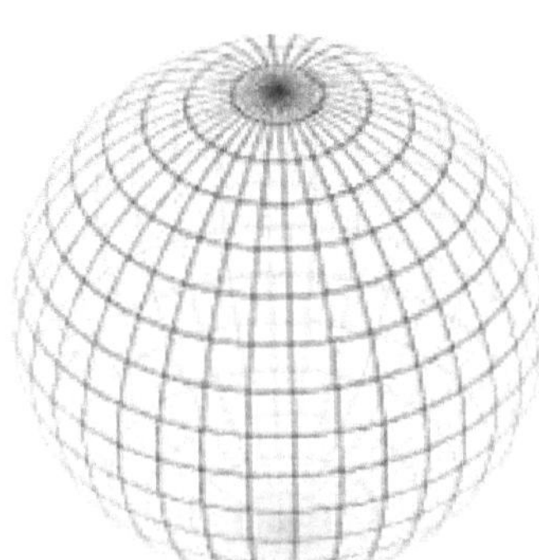

A sphere is a surface that can be defined parametrically (by x = r sin è cos ö, y = r sin è sin ö, z= r cos è) or implicitly (by x + y + z ″ r = 0.)

A surface is a two-dimensional object, such as a sphere or paraboloid. In differential geometry and topology, surfaces are described by two-dimensional 'patches' (or neighborhoods) that are assembled by diffeomorphisms or homeomorphisms, respectively. In algebraic geometry, surfaces are described by polynomial equations.

Manifolds

A manifold is a generalization of the concepts of curve and surface. In topology, a manifold is a topological space where every point has a neighborhoodthat is homeomorphic to Euclidean space. In differential geometry, a differentiable manifold is a space where each neighborhood is diffeomorphic to Euclidean space.

Manifolds are used extensively in physics, including in general relativity and string theory

Topologies and metrics\

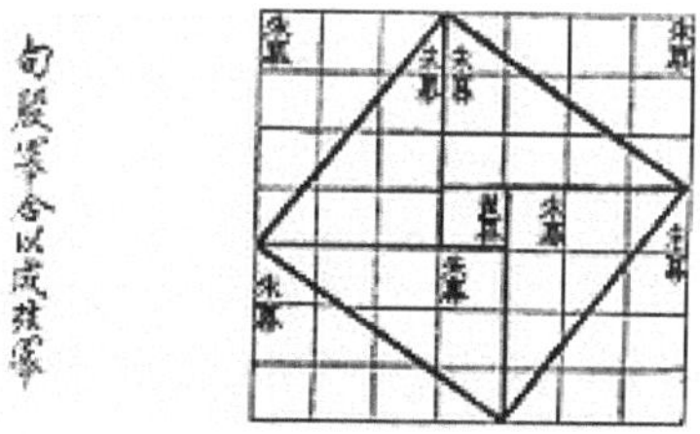

Visual checking of the Pythagorean theorem for the (3, 4, 5) triangle as in the Zhoubi Suanjing 500–200 BC. The Pythagorean theorem is a consequence of the Euclidean metric.

A topology is a mathematical structure on a set that tells how elements of the set relate spatially to each other. The best-known examples of topologies come from metrics, which are ways of measuring distances between points. For instance, the Euclidean metric measures the distance between points in the Euclidean plane, while the hyperbolic metric measures the distance in the hyperbolic plane. Other important examples of metrics include the Lorentz metric of special relativity and the semi-Riemannian metrics of general relativity.

Compass and straightedge constructions

Classical geometers paid special attention to constructing geometric objects that had been described in some other way. Classically, the only instruments allowed in geometric constructions are the compass and straightedge. Also, every construction had to be complete in a finite number of steps. However, some problems turned out to be difficult or impossible to solve by these means alone, and ingenious constructions using parabolas and other curves, as well as mechanical devices, were found.

Dimension

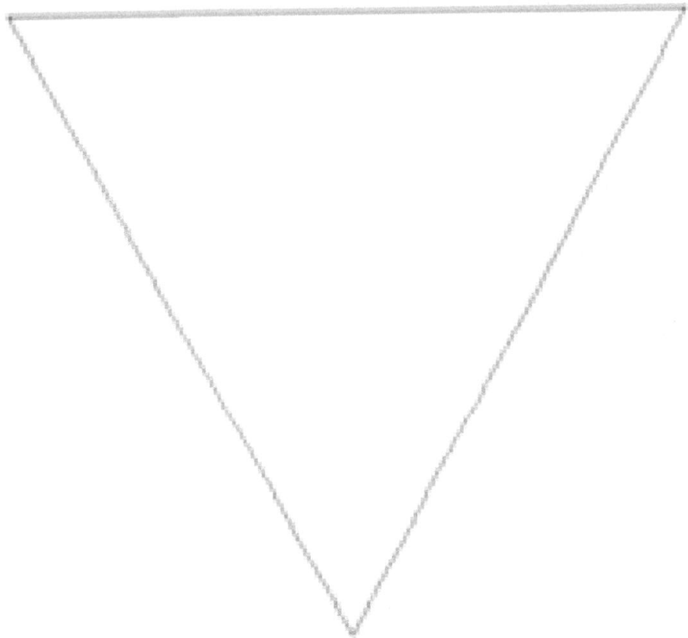

The Koch snowflake, with fractal dimension=log4/log3 and topological dimension=1

Where the traditional geometry allowed dimensions 1 (a line), 2 (a plane) and 3 (our ambient world conceived of as three-dimensional space), mathematicians have used higher dimensions for nearly two centuries. The concept of dimension has gone through stages of being any natural number n, to being possibly infinite with the introduction of Hilbert space, to being any positive real number in fractal geometry. Dimension theory is a technical area, initially within general topology, that discusses definitions; in common with most mathematical ideas, dimension is now defined rather than an intuition. Connected topological manifolds have a well-defined dimension; this is a theorem (invariance of domain) rather than anything a priori.

The issue of dimension still matters to geometry as many classic questions still lack complete answers. For instance, many open problems in topology depend on the dimension of an object for the result. In physics, dimensions 3 of space and 4 of space-time are special cases in geometric topology, and dimensions 10 and 11 are key ideas in string theory. Currently, the existence of the theoretical dimensions is purely defined by technical reasons; it is likely that further research may result in a geometric reason for the significance of 10 or 11 dimensions in the theory, lending credibility or possibly disproving string theory.

Symmetry

A tiling of the hyperbolic plane

The theme of symmetry in geometry is nearly as old as the science of geometry itself. Symmetric shapes such as the circle, regular polygonsand platonic solids held deep significance for many ancient philosophers and were investigated in detail before the time of Euclid. Symmetric patterns occur in nature and were artistically rendered in a multitude of forms, including the graphics of M.C. Escher. Nonetheless, it was not until the second half of 19th century that the unifying role of symmetry in foundations of geometry was recognized. Felix Klein's Erlangen program proclaimed that, in a very precise sense, symmetry, expressed via the notion of a transformation group, determines what geometry is. Symmetry in classical Euclidean geometry is represented by congruences and rigid motions, whereas in projective geometry an analogous role is played by collineations, geometric transformations that take straight lines into straight lines. However it was in the new geometries of Bolyai and Lobachevsky, Riemann, Clifford and Klein, and Sophus Lie that Klein's idea to 'define a geometry via its symmetry group' proved most influential. Both discrete and continuous symmetries play prominent roles in geometry, the former in topology and geometric group theory, the latter in Lie theory and Riemannian geometry.

A different type of symmetry is the principle of duality in projective geometry (see Duality (projective geometry)) among other fields. This meta-phenomenon can roughly be described as follows: in any theorem, exchange point with plane, join with meet, lies in with contains, and you will get an equally true theorem. A similar and closely related form of duality exists between a vector space and its dual space.

Non-Euclidean geometry

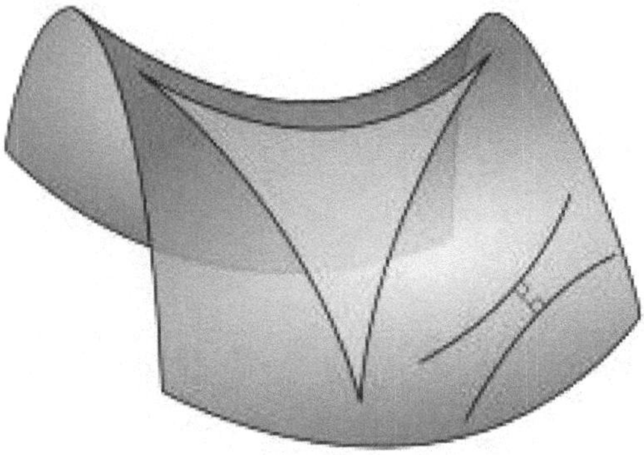

Differential geometry uses tools from calculus to study problems involving curvature.

In the nearly two thousand years since Euclid, while the range of geometrical questions asked and answered inevitably expanded, the basic understanding of space remained essentially the same. Immanuel Kant argued that there is only one, absolute, geometry, which is known to be true a priori by an inner faculty of mind: Euclidean geometry was synthetic a priori. This dominant view was overturned by the revolutionary discovery of non-Euclidean geometry in the works of Bolyai, Lobachevsky, and Gauss (who never published his theory). They demonstrated that ordinary Euclidean space is only one possibility for development of geometry. A broad vision of the subject of geometry was then expressed by Riemann in his 1867 inauguration lecture Über die Hypothesen, welche der Geometrie zu Grunde liegen (On the hypotheses on which geometry is based), published only after his death. Riemann's new idea of space proved crucial in Einstein's general relativity theory, and Riemannian geometry, that considers very general spaces in which the notion of length is defined, is a mainstay of modern geometry.

Contemporary geometry

Euclidean geometry

Euclidean geometry has become closely connected with computational geometry, computer graphics, convex geometry, incidence geometry, finite geometry, discrete geometry, and some areas of combinatorics. Attention was given to further work on Euclidean geometry and the Euclidean groups by crystallography and the work of H. S. M. Coxeter, and can be seen in theories of Coxeter groups and polytopes. Geometric group theory is an expanding area of the theory of more general discrete groups, drawing on geometric models and algebraic techniques.

Differential geometry

Differential geometry has been of increasing importance to mathematical physics due to Einstein's general relativity postulation that the universe is curved. Contemporary differential geometry is intrinsic, meaning that the spaces it considers are smooth manifolds whose geometric structure is governed by a Riemannian metric, which determines how distances are measured near each point, and not a prioriparts of some ambient flat Euclidean space.

Topology and geometry

The field of topology, which saw massive development in the 20th century, is in a technical sense a type of transformation geometry, in which transformations are homeomorphisms. This has often been expressed in the form of the dictum 'topology is rubber-sheet geometry'. Contemporary geometric topology and differential topology, and particular subfields such as Morse theory, would be counted by most mathematicians as part of geometry. Algebraic topology and general topology have gone their own ways.

Algebraic geometry

The field of algebraic geometry is the modern incarnation of the Cartesian geometry of co-ordinates. From late 1950s through mid-1970s it had undergone major foundational development, largely due to work of Jean-Pierre Serre and Alexander Grothendieck. This led to the introduction of schemes and greater emphasis on topological methods, including various cohomology theories. One of seven Millennium Prize problems, the Hodge conjecture, is a question in algebraic geometry.

The study of low-dimensional algebraic varieties, algebraic curves, algebraic surfaces and algebraic varieties of dimension 3 ("algebraic threefolds"), has been far advanced. Gröbner basis theory and real algebraic geometry are among more applied subfields of modern algebraic geometry. Arithmetic geometryis an active field combining algebraic geometry and number theory. Other directions of research involve moduli spaces and complex geometry. Algebro-geometric methods are commonly applied in string and brane theory.

Applications

Geometry has found applications in many fields, some of which are described below.

Art

Mathematics and art are related in a variety of ways. For instance, the theory of perspective showed that there is more to geometry than just the metric properties of figures: perspective is the origin of projective geometry.

Architecture

Mathematics and architecture are related, since, as with other arts, architects use mathematics for several reasons. Apart from the mathematics needed when engineering buildings, architects use geometry: to define the spatial form of a building; from the Pythagoreans of the sixth century BC onwards, to create forms considered harmonious, and thus to lay out buildings and their surroundings according to mathematical, aesthetic and sometimes religious principles; to decorate buildings with mathematical objects such as tessellations; and to meet environmental goals, such as to minimise wind speeds around the bases of tall buildings.

Physics

The 4polytope, orthogonally projected into the ELie group Coxeter plane. Lie groups have several applications in physics.

The field of astronomy, especially as it relates to mapping the positions of stars and planets on the celestial sphere and describing the relationship between movements of celestial bodies, have served as an important source of geometric problems throughout history.

Modern geometry has many ties to physics as is exemplified by the links between pseudo-Riemannian geometry and general relativity. Onc of the youngest physical theories, string theory, is also very geometric in flavour.

Other fields of mathematics

Geometry has also had a large effect on other areas of mathematics. For instance, the introduction of coordinates by René Descartes and the concurrent developments of algebra marked a new stage for geometry, since geometric figures such as plane curves could now be represented analytically in the form of functions and equations. This played a key role in the emergence of infinitesimal calculus in the 17th century. The subject of geometry was further enriched by the study of the intrinsic structure of geometric objects that originated with Euler and Gauss and led to the creation of topology and differential geometry.

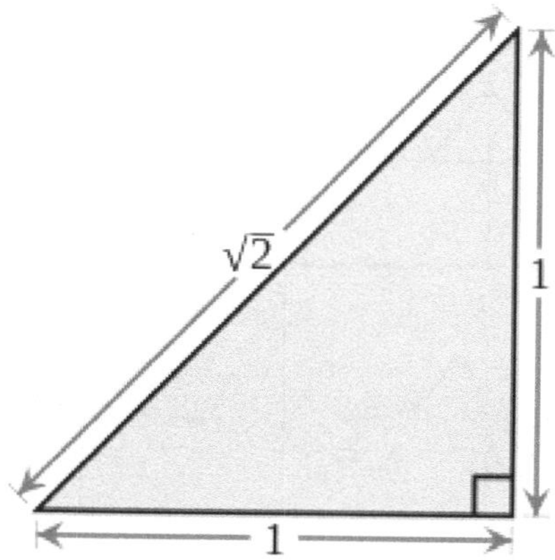

The Pythagoreans discovered that the sides of a triangle could have incommensurable lengths.

An important area of application is number theory. In ancient Greece the Pythagoreans considered the role of numbers in geometry. However, the discovery of incommensurable lengths, which contradicted their philosophical views, made them abandon abstract numbers in favor of concrete geometric quantities, such as length and area of figures. Since the 19th century, geometry has been used for solving problems in number theory, for example through the geometry of numbers or, more recently, scheme theory, which is used in Wiles's proof of Fermat's Last Theorem.

While the visual nature of geometry makes it initially more accessible than other mathematical areas such as algebra or number theory, geometric language is also used in contexts far removed from its traditional, Euclidean provenance (for example, in fractal geometry and algebraic geometry).

Analytic geometry applies methods of algebra to geometric questions, typically by relating geometric curves to algebraic equations. These ideas played a key role in the development of calculus in the 17th century and led to the discovery of many new properties of plane curves. Modern algebraic geometry considers similar questions on a vastly more abstract level.

Leonhard Euler, in studying problems like the Seven Bridges of Königsberg, considered the most fundamental properties of geometric figures based solely on shape, independent of their metric properties. Euler called this new branch of geometry geometria situs (geometry of place), but it is now known as topology. Topology grew out of geometry, but turned into a large independent discipline. It does not differentiate between objects that can be continuously deformed into each other. The objects may nevertheless retain some geometry, as in the case of hyperbolic knots.

Mathematical physics

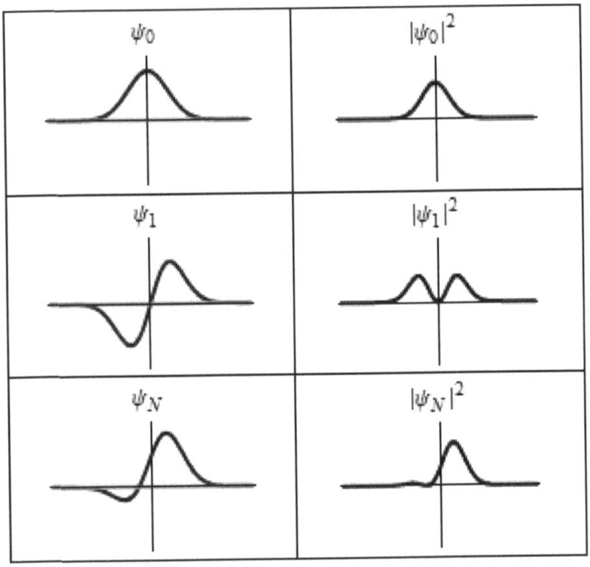

An example of mathematical physics: solutions of Schrödinger's equation for quantum harmonic oscillators (left) with their amplitudes (right).

Mathematical physics refers to the development of mathematical methods for application to problems in physics. The Journal of Mathematical Physics defines the field as "the application of mathematics to problems in physics and the development of mathematical methods suitable for such applications and for the formulation of physical theories". It is a branch of applied mathematics, but deals with physical problems.

Scope

There are several distinct branches of mathematical physics, and these roughly correspond to particular historical periods.

Classical mechanics

The rigorous, abstract and advanced reformulation of Newtonian mechanics adopting the Lagrangian mechanics and the Hamiltonian mechanics even in the presence of constraints. Both formulations are embodied in analytical mechanics. It leads, for instance, to discover the deep interplay of the notion of symmetry and that of conserved quantities during the dynamical evolution, stated within the most elementary formulation of Noether's theorem. These approaches and ideas can be, and in fact have been, extended to other areas of physics as statistical mechanics, continuum mechanics, classical field theory and quantum field theory. Moreover, they have provided several examples and basic ideas in differential geometry (e.g. the theory of vector bundles and several notions in symplectic geometry).

Partial differential equations

The theory of partial differential equations (and the related areas of variational calculus, Fourier analysis, potential theory, and vector analysis) are perhaps most closely associated with mathematical physics. These were developed intensively from the second half of the 18th century (by, for example, D'Alembert, Euler, and Lagrange) until the 1930s. Physical applications of these developments include hydrodynamics, celestial mechanics, continuum mechanics, elasticity theory, acoustics, thermodynamics, electricity, magnetism, and aerodynamics.

Quantum theory

The theory of atomic spectra (and, later, quantum mechanics) developed almost concurrently with the mathematical fields of linear algebra, the spectral theory of operators, operator algebras and more broadly, functional analysis. Nonrelativistic quantum mechanics includes Schrödinger operators, and it has connections to atomic and molecular physics. Quantum information theory is another subspecialty.

RELATIVITY AND QUANTUM RELATIVISTIC THEORIES

The special and general theories of relativity require a rather different type of mathematics. This was group theory, which played an important role in both quantum field theory and differential geometry. This was, however, gradually supplemented by topology and functional analysis in the mathematical description of cosmological as well as quantum field theory phenomena. In this area both homological algebra and category theory are important nowadays.

Statistical mechanics

Statistical mechanics forms a separate field, which includes the theory of phase transitions. It relies upon the Hamiltonian mechanics (or its quantum version) and it is closely related with the more mathematical ergodic theory and some parts of probability theory. There are increasing interactions between combinatorics and physics, in particular statistical physics.

Usage

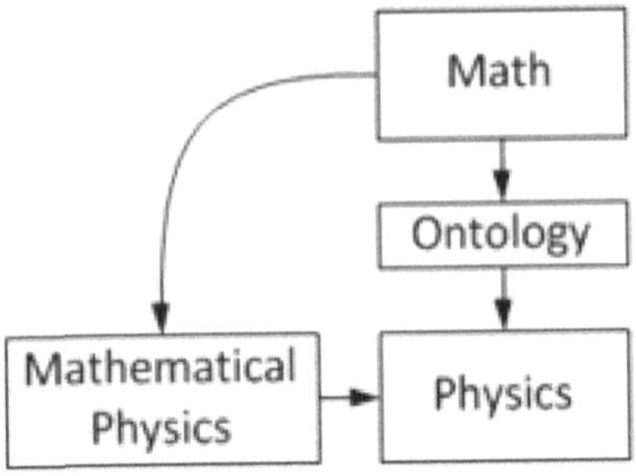

The usage of the term "mathematical physics" is sometimes idiosyncratic. Certain parts of mathematics that initially arose from the development of physics are not, in fact, considered parts of mathematical physics, while other closely related fields are. For example, ordinary differential equations and symplectic geometry are generally viewed as purely mathematical disciplines, whereas dynamical systemsand Hamiltonian mechanics belong to mathematical physics. John Herapath used the term for the title of his 1847 text on "mathematical principles of natural philosophy"; the scope at that time being "the causes of heat, gaseous elasticity, gravitation, and other great phenomena of nature".

Mathematical vs. theoretical physics

The term "mathematical physics" is sometimes used to denote research aimed at studying and solving problems inspired by physics or thought experiments within a mathematically rigorous framework. In this sense, mathematical physics covers a very broad academic realm distinguished only by the blending of pure mathematics and physics. Although related to theoretical physics, mathematical physics in this sense emphasizes the mathematical rigour of the same type as found in mathematics.

On the other hand, theoretical physics emphasizes the links to observations and experimental physics, which often requires theoretical physicists (and mathematical physicists in the more general sense) to use heuristic, intuitive, and approximate arguments. Such arguments are not considered rigorous by mathematicians, but that is changing over time.

Such mathematical physicists primarily expand and elucidate physical theories. Because of the required level of mathematical rigour, these researchers often deal with questions that theoretical physicists have considered to be already solved. However, they can sometimes show (but neither commonly nor easily) that the previous solution was incomplete, incorrect, or simply too naïve. Issues about attempts to infer the second law of thermodynamics from statistical mechanics are examples. Other examples concern the subtleties involved with synchronisation procedures in special and general relativity (Sagnac effect and Einstein synchronisation).

The effort to put physical theories on a mathematically rigorous footing has inspired many mathematical developments. For example, the development of quantum mechanics and some aspects of functional analysis parallel each other in many ways. The mathematical study of quantum mechanics, quantum field theory, and quantum statistical mechanics has motivated results in operator algebras. The attempt to construct a rigorous quantum field theory has also brought about progress in fields such as representation theory. Use of geometry and topology plays an important role in string theory.

Prominent mathematical physicists

Before Newton

The roots of mathematical physics can be traced back to the likes of Archimedes in Greece, Ptolemy in Egypt, Alhazen in Iraq, and Al-Biruni in Persia.

In the first decade of the 16th century, amateur astronomer Nicolaus Copernicus proposed heliocentrism, and published a treatise on it in 1543. He retained the Ptolemaic idea of epicycles, and merely sought to simplify astronomy by constructing simpler sets of epicyclic orbits. Epicycles consist of circles upon circles. According to Aristotelian physics, the circle was the perfect form of motion, and was the intrinsic motion of Aristotle's fifth element—the quintessence or universal essence known in Greek as aether for the English pure air—that was the pure substance beyond the sublunary sphere, and thus was celestial entities' pure composition. The German Johannes Kepler [1571–1630], Tycho Brahe's assistant, modified Copernican orbits to ellipses, formalized in the equations of Kepler's laws of planetary motion.

An enthusiastic atomist, Galileo Galilei in his 1623 book The Assayer asserted that the "book of nature" is written in mathematics. His 1632 book, about his telescopic observations, supported heliocentrism. Having introduced experimentation, Galileo then refuted geocentric cosmology by refuting Aristotelian physics itself. Galilei's 1638 book Discourse on Two New Sciences established the law of equal free fall as well as the principles of inertial motion, founding the central concepts of what would become today's classical mechanics. By the Galilean law of inertia as well as the principle of Galilean invariance, also called Galilean relativity, for any object experiencing inertia, there is empirical justification for knowing only that it is at relative rest or relative motion—rest or motion with respect to another object.

René Descartes adopted Galilean principles and developed a complete system of heliocentric cosmology, anchored on the principle of vortex motion, Cartesian physics, whose widespread acceptance brought the demise of Aristotelian physics. Descartes sought to formalize mathematical reasoning in science, and developed Cartesian coordinates for geometrically plotting locations in 3D space and marking their progressions along the flow of time.

Christiaan Huygens was the first to use mathematical formulas to describe the laws of physics, and for that reason Huygens is regarded as the first theoretical physicist and the founder of mathematical physics.

Newtonian and post Newtonian

Isaac Newton (1642–1727) developed new mathematics, including calculus and several numerical methods such as Newton's method to solve problems in physics. Newton's theory of motion, published in 1687, modeled three Galilean laws of motion along with Newton's law of universal gravitation on a framework of absolute space—hypothesized by Newton as a physically real entity of Euclidean geometric structure extending infinitely in all directions—while presuming absolute time, supposedly justifying knowledge of absolute motion, the object's motion with respect to absolute space. The principle of Galilean invariance/relativity was merely implicit in Newton's theory of motion. Having ostensibly reduced the Keplerian celestial laws of motion as well as Galilean terrestrial laws of motion to a unifying force, Newton achieved great mathematical rigor, but with theoretical laxity.

In the 18th century, the Swiss Daniel Bernoulli (1700–1782) made contributions to fluid dynamics, and vibrating strings. The Swiss Leonhard Euler (1707–1783) did special work in variational calculus, dynamics, fluid dynamics, and other areas. Also notable was the Italian-born Frenchman, Joseph-Louis Lagrange (1736–1813) for work in analytical mechanics: he formulated Lagrangian mechanics) and variational methods. A major contribution to the formulation of Analytical Dynamics called Hamiltonian dynamics was also made by the Irish physicist, astronomer and mathematician, William Rowan Hamilton (1805-1865). Hamiltonian dynamics had played an important role in the formulation of modern theories in physics, including field theory and quantum mechanics. The French mathematical physicist Joseph Fourier (1768 – 1830) introduced the notion of Fourier series to solve the heat equation, giving rise to a new approach to solving partial differential equations by means of integral transforms.

Into the early 19th century, the French Pierre-Simon Laplace (1749–1827) made paramount contributions to mathematical astronomy, potential theory, and probability theory. Siméon Denis Poisson (1781–1840) worked in analytical mechanics and potential theory. In Germany, Carl Friedrich Gauss (1777–1855) made key contributions to the theoretical foundations of electricity, magnetism, mechanics, and fluid dynamics. In England, George Green (1793-1841) published An Essay on the Application of Mathematical Analysis to the Theories of Electricity and Magnetism in 1828, which in addition to its significant contributions to mathematics made early progress towards laying down the mathematical foundations of electricity and magnetism.

A couple of decades ahead of Newton's publication of a particle theory of light, the Dutch Christiaan Huygens (1629–1695) developed the wave theory of light, published in 1690. By 1804, Thomas Young's double-slit experiment revealed an interference pattern, as though light were a wave, and thus Huygens's wave theory of light, as well as Huygens's inference that light waves were vibrations of the luminiferous aether, was accepted. Jean-Augustin Fresnel modeled hypothetical behavior of the aether. Michael Faraday introduced the theoretical concept of a field—not action at a distance. Mid-19th century, the Scottish James Clerk Maxwell (1831–1879) reduced electricity and magnetism to Maxwell's electromagnetic field theory, whittled down by others to the four Maxwell's equations. Initially, optics was found consequent of Maxwell's field. Later, radiation and then today's known electromagnetic spectrum were found also consequent of this electromagnetic field.

The English physicist Lord Rayleigh [1842–1919] worked on sound. The Irishmen William Rowan Hamilton (1805–1865), George Gabriel Stokes (1819–1903) and Lord Kelvin(1824–1907) produced several major works: Stokes was a leader in optics and fluid dynamics; Kelvin made substantial discoveries in thermodynamics; Hamilton did notable work on analytical mechanics, discovering a new and powerful approach nowadays known as Hamiltonian mechanics. Very relevant contributions to this approach are due to his German colleague Carl Gustav Jacobi (1804–1851) in particular referring to canonical transformations. The German Hermann von Helmholtz (1821–1894) made substantial contributions in the fields of electromagnetism, waves, fluids, and sound. In the United States, the pioneering work of Josiah Willard Gibbs (1839–1903) became the basis for statistical mechanics. Fundamental theoretical results in this area were achieved by the German Ludwig Boltzmann (1844-1906). Together, these individuals laid the foundations of electromagnetic theory, fluid dynamics, and statistical mechanics.

Relativistic

By the 1880s, there was a prominent paradox that an observer within Maxwell's electromagnetic field measured it at approximately constant speed, regardless of the observer's speed relative to other objects within the electromagnetic field. Thus, although the observer's speed was continually lost relative to the electromagnetic field, it was preserved relative to other objects in the electromagnetic field. And yet no violation of Galilean invariance within physical interactions among objects was detected. As Maxwell's electromagnetic field was modeled as oscillations of the aether, physicists inferred that motion within the aether resulted in aether drift, shifting the electromagnetic field, explaining the observer's missing speed relative to it. The Galilean transformation had been the mathematical process used to translate the positions in one reference frame to predictions of positions in another reference frame, all plotted on Cartesian coordinates, but this process was replaced by Lorentz transformation, modeled by the Dutch Hendrik Lorentz [1853–1928].

In 1887, experimentalists Michelson and Morley failed to detect aether drift, however. It was hypothesized that motion into the aether prompted aether's shortening, too, as modeled in the Lorentz contraction. It was hypothesized that the aether thus kept Maxwell's electromagnetic field aligned with the principle of Galilean invariance across all inertial frames of reference, while Newton's theory of motion was spared.

In the 19th century, Gauss's contributions to non-Euclidean geometry, or geometry on curved surfaces, laid the groundwork for the subsequent development of Riemannian geometryby Bernhard Riemann (1826–1866). Austrian theoretical physicist and philosopher Ernst Mach criticized Newton's postulated absolute space. Mathematician Jules-Henri Poincaré(1854–1912) questioned even absolute time. In 1905, Pierre Duhem published a devastating criticism of the foundation of Newton's theory of motion. Also in 1905, Albert Einstein(1879–1955) published his special theory of relativity, newly explaining both the electromagnetic field's invariance and Galilean invariance by discarding all hypotheses concerning aether, including the existence of aether itself. Refuting the framework of Newton's theory—absolute space and absolute time—special relativity refers to relative space and relative time, whereby length contracts and time dilates along the travel pathway of an object.

In 1908, Einstein's former professor Hermann Minkowski modeled 3D space together with the 1D axis of time by treating the temporal axis like a fourth spatial dimension—altogether 4D spacetime—and declared the imminent demise of the separation of space and time. Einstein initially called this "superfluous learnedness", but later used Minkowski spacetimewith great elegance in his general theory of relativity, extending invariance to all reference frames—whether perceived as inertial or as accelerated—and credited this to Minkowski, by then deceased. General relativity replaces Cartesian coordinates with Gaussian coordinates, and replaces Newton's claimed empty yet Euclidean space traversed instantly by Newton's vector of hypothetical gravitational force—an instant action at a distance—with a gravitational field. The gravitational field is Minkowski spacetime itself, the 4D topology of Einstein aether modeled on a Lorentzian manifold that "curves" geometrically, according to the Riemann curvature tensor, in the vicinity of either mass or energy. (Under special relativity—a special case of general relativity—even massless energy exerts gravitational effect by its mass equivalence locally "curving" the geometry of the four, unified dimensions of space and time.)

QUANTUM

Another revolutionary development of the 20th century was quantum theory, which emerged from the seminal contributions of Max Planck (1856–1947) (on black body radiation) and Einstein's work on the photoelectric effect. This was, at first, followed by a heuristic framework devised by Arnold Sommerfeld (1868–1951) and Niels Bohr (1885–1962), but this was soon replaced by the quantum mechanics developed by Max Born (1882–1970), Werner Heisenberg (1901–1976), Paul Dirac (1902–1984), Erwin Schrödinger (1887–1961), Satyendra Nath Bose (1894–1974), and Wolfgang Pauli (1900–1958). This revolutionary theoretical framework is based on a probabilistic interpretation of states, and evolution and measurements in terms of self-adjoint operators on an infinite dimensional vector space. That is called Hilbert space, introduced in its elementary form by David Hilbert (1862–1943) and Frigyes Riesz (1880-1956), and rigorously defined within the axiomatic modern version by John von Neumann in his celebrated book Mathematical Foundations of Quantum Mechanics, where he built up a relevant part of modern functional analysis on Hilbert spaces, the spectral theory in particular. Paul Dirac used algebraic constructions to produce a relativistic model for the electron, predicting its magnetic moment and the existence of its antiparticle, the positron.

Architectural Geometry

Architectural geometry is an area of research which combines applied geometry and architecture, which looks at the design, analysis and manufacture processes. It lies at the core of architectural design and strongly challenges contemporary practice, the so-called architectural practice of the digital age.

Architectural geometry is influenced by following fields: differential geometry, topology, fractal geometry, and cellular automata.

Topics include:

- Freeform curves and surfaces creation
- Developable surfaces
- Discretisation
- Generative design
- Digital prototyping and manufacturing

9

Coordinate Geometry

In geometry, a coordinate system is a system which uses one or more numbers, or coordinates, to uniquely determine the position of a point or other geometric element on a manifold such asEuclidean space. The order of the coordinates is significant and they are sometimes identified by their position in an ordered tuple and sometimes by a letter, as in 'the x-coordinate'. The coordinates are taken to be real numbers in elementary mathematics, but may be complex numbers or elements of a more abstract system such as a commutative ring. The use of a coordinate system allows problems in geometry to be translated into problems about numbers and vice versa; this is the basis of analytic geometry.

A system of geometry where the position of points on the plane is described using an ordered pair of numbers.

Recall that a plane is a flat surface that goes on forever in both directions. If we were to place a point on the plane, coordinate geometry gives us a way to describe exactly where it is by using two numbers.

To introduce the idea, consider the grid on the right. The columns of the grid are lettered A,B,C etc. The rows are numbered 1,2,3 etc from the top. We can see that the X is in box D3; that is, column D, row 3.

	A	B	C	D	E	F
1						
2						
3				X		
4						
5						
6						

D and 3 are called the coordinates of the box. It has two parts: the row and the column. There are many boxes in each row and many boxes in each column. But by having both we can find one single box, where the row and column intersect.

3D Basics

3D models are stored using a 3D coordinate system. This page describes some of the more common elements of a coordinate system that you will need to be familiar with when modeling.

Coordinates

Every point of a 3D model is mapped to a location in space measured along X, Y and Z. When taken together, {X,Y,Z} is called a coordinate. Wings 3D uses the Cartesian coordinate system.

Axes XYZ

Looking at the Wings 3D workspace, you can see the red, blue, and green lines that represent the X,Y, and Z axes.

In Wings 3D, Y is up and down, X is side to side, and Z is front to back. These three axes are commonly used to define the active axis for moving, scaling, or rotating a model. Often in Wings 3D they will be referred to as Standard axes.

Note: Other modelers may swap the Y and Z axis, so when importing a model into Wings 3D it may appear to be the wrong way around.

Points

A point is any location in 3D space. Points can be used as references, such as the center around which a model will be rotated. A point may be defined by the intersection of two vectors. Points can also be used to define the location of a plane.

Vectors

A vector is defined by two points. Adding the coordinates of two points results in a vector coordinate including its magnitude.

A vector is generally necessary to establish a direction or axis depending on the context in which the vector is to be used. A vector is often used to define a direction for movement or an axis of rotation.

Normals

An un-normalised vector denotes not just direction but also magnitude. Normalised vectors have a magnitude of 1.0, hence the meaning of normalisation.

Planes

A plane is defined by a minimum of three unique points in space. These points make a triangle, which is the minimum requirement for defining a surface. Once we have these points, they can be added together to determine the plane vector. A plane's location can be determined by finding the center or average of its defining points. However, if a plane's vector is normalised, it represents a surface located anywhere in space. A normalised plane will often be used in conjunction with a point to place the plane.

To picture a plane vector, think of a table. The surface of the table represents a plane along which things can move in two directions. The plane vector as represented as a coordinate is actually a vector point straight up or perpendicular to the table's surface. If we normalise this vector, then we can relocate the surface of the table to any parallel plane in the room. But once the table in rotated, the plane has changed. All parallel planes are represented by the same plane normal.

THE COORDINATE PLANE

In coordinate geometry, points are placed on the "coordinate plane" as shown below. It has two scales - one running across the plane called the "x axis" and another a right angles to it called the y axis. (These can be thought of as similar to the column and row in the paragraph above.) The point where the axes cross is called the origin and is where both x and y are zero.

The coordinate plane is a two-dimensional surface on which we can plot points, lines and curves. It has two scales, called the x-axis and y-axis, at right angles to each other. The plural of axis is 'axes' (pronounced "AXE-ease"). Points on the plane are located using two numbers - thye x and y coordinates. These are the horizontal and vertical distances of the point from a specific location called the origin.

X axis

The horizontal scale is called the x-axis. As you go to the right on the scale from zero, the values are positive and get larger. As you go to the left from zero, they get more and more negative.

Y axis

The vertical scale is called the y-axis. As you go up from zero the numbers are increasing in a positive direction. As you go down from zero they get more and more negative.

Axis labelling

Along each axis you will see small tic marks with numbers. These are labels to help judge the scale. They are shown every 5 units in the figure above, but can be any increment, and need not be the same on both axes.

Origin

The point where the two axes cross (at zero on both scales) is called the origin. The origin is the point from which all distances along the x and y axes are measured. In the figure above you can drag the origin point to reposition it to a more suitable location at any time.

Quadrants

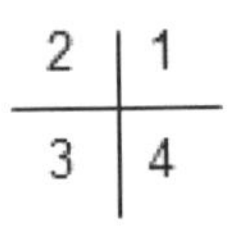

The two axes divide the plane into four areas called quadrants. The first quadrant, by convention, is the top right, and then they go around counter-clockwise. In the diagram above they are labelled Quadrant 1,2 etc. It is conventional to label them with numerals but we talk about them as "first, second, third, and fourth quadrant". They are also sometimes labelled with Roman numerals: I, II, III and IV.

Points and Coordinates

Every point in space can be assigned three numbers with respect to a starting point. Those three numbers allow us to distinguish any point from any other point in space. Fortunately for you, we are not dealing here with three dimensions, but only with two.

- Coordinates of a point: Each point on a number line is assigned a number. In the same way, each point in a plane is assigned a pair of numbers.
- x-axis and y-axis: To locate points in a plane, two perpendicular lines are used—a horizontal line called the x-axis and a vertical line called the y-axis.
- Origin: The point of intersection of the x-axis and y-axis.
- Coordinate plane: The x-axis, y-axis, and all the points in the plane they determine.
- Ordered pairs: Every point in a coordinate plane is named by a pair of numbers whose order is important; these numbers are written in parentheses and separated by a comma.
- x-coordinate : The number to the left of the comma in an ordered pair is thex-coordinate of the point and indicates the amount of movement

along the x-axis from the origin. The movement is to the right if the number is positive and to the left if the number is negative.

- y-coordinate : The number to the right of the comma in an ordered pair is the y-coordinate of the point and indicates the amount of movement perpendicular to the x-axis. The movement is above the x-axis if the number is positive and below the x-axis if the number is negative.
  Note: The coordinates [ordered pair] for the origin are (0, 0).

The x-axis and y-axis separate the coordinate plane into four regions calledquadrants. (See Figure 1 .) The upper right quadrant is quadrant 1; the upper left quadrant is quadrant II; the lower left quadrant is quadrant III; and the lower right quadrant is quadrant IV. Notice the following:

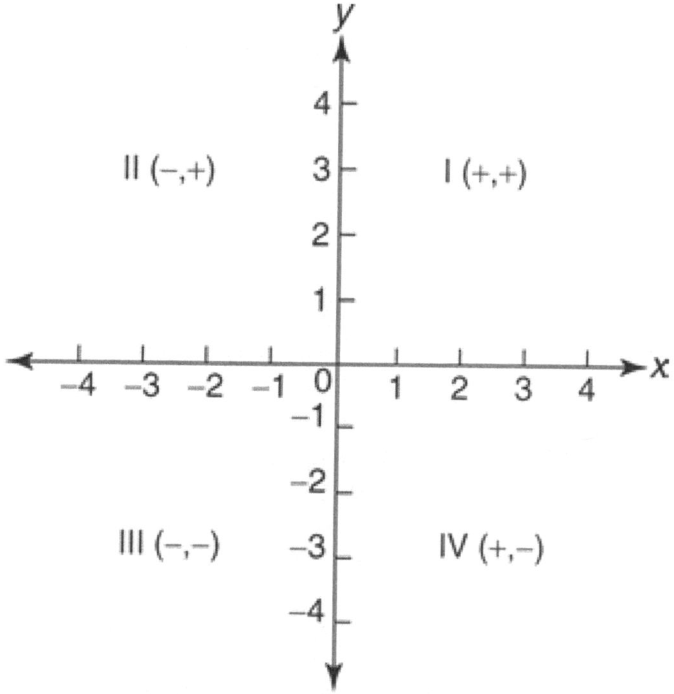

The coordinate axes separate the plane into four quadrants.

- In quadrant I, x is always positive and y is always positive.
- In quadrant II, x is always negative and y is always positive.
- In quadrant III, x is always negative and y is always negative.
- In quadrant IV, x is always positive and y is always negative.

The point associated with an ordered pair of real numbers is called the graph of the ordered pair.

Distance Formula

In Figure 1 , A is (2, 2), B is (5, 2), and C is (5, 6)

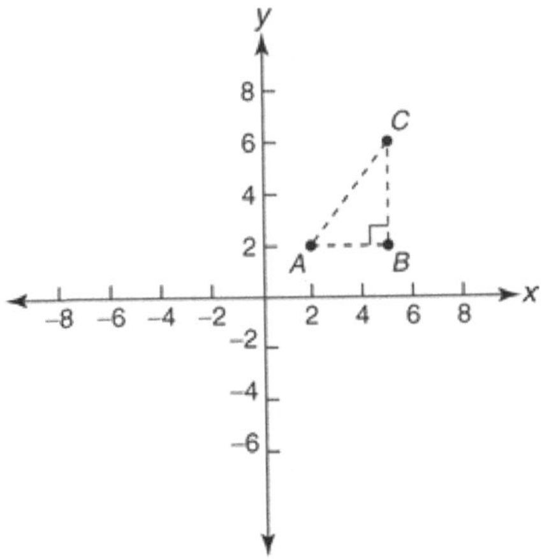

Figure 1 Finding the distance from A to C.

$AB = 5 - 2$	and	$BC = 6 - 2$
$AB = 3$		$BC = 4$

$$\overline{AC}^2 = AB^2 + BC^2$$
$$\overline{AC} = \sqrt{AB^2 + BC^2}$$
$$\overline{AC} = \sqrt{3^2 + 4^2}$$
$$\overline{AC} = \sqrt{9 + 16}$$
$$\overline{AC} = \sqrt{25}$$
$$\overline{AC} = 5$$

If A is represented by the ordered pair (x_1, y_1) and C is represented by the ordered pair (x_2, y_2), then AB = (x_2 ″ x_1) and BC = (y_2 ″ y_1).

Then

$$AC = \sqrt{(x_2 - x_1)^2 + (y_2 - y_1)^2}$$

This is stated as a theorem.

This is stated as a theorem.

Theorem 101: If the coordinates of two points are (x_1, y_1) and (x_2, y_2), then the distance, d, between the two points is given by the following formula (Distance Formula).

$$d = \sqrt{(x_2 - x_1)^2 + (y_2 - y_1)^2}$$

Example 1: Use the Distance Formula to find the distance between the points with coordinates (“3, 4) and (5, 2).

Let (–3, 4) = (x_1, y_1) and (5, 2) = (x_2, y_2). Then

$$d = \sqrt{(5-(-3))^2+(2-4)^2}$$
$$d = \sqrt{(8)^2+(2)^2}$$
$$d = \sqrt{64+4}$$
$$d = \sqrt{68}$$
$$d = \sqrt{(4)(17)}$$
$$d = 2\sqrt{17}$$

Example 2: A triangle has vertices A(12,5), B(5,3), and C(12, 1). Show that the triangle is isosceles.

By the Distance Formula,

$$AB = \sqrt{(5-12)^2+(3-5)^2} \qquad BC = \sqrt{(12-5)^2+(1-3)^2}$$
$$AB = \sqrt{(7^2)+(-2)} \qquad BC = \sqrt{7^2+(-2)^2}$$
$$AB = \sqrt{49+4} \qquad BC = \sqrt{49+4}$$
$$AB = \sqrt{53} \qquad BC = \sqrt{53}$$

Because AB = BC, triangle ABC is isosceles.

Coordinate Space

Before we begin programming with Processing, we must first channel our eighth grade selves, pull out a piece of graph paper, and draw a line. The shortest distance between two points is a good old fashioned line, and this is where we begin, with two points on that graph paper.

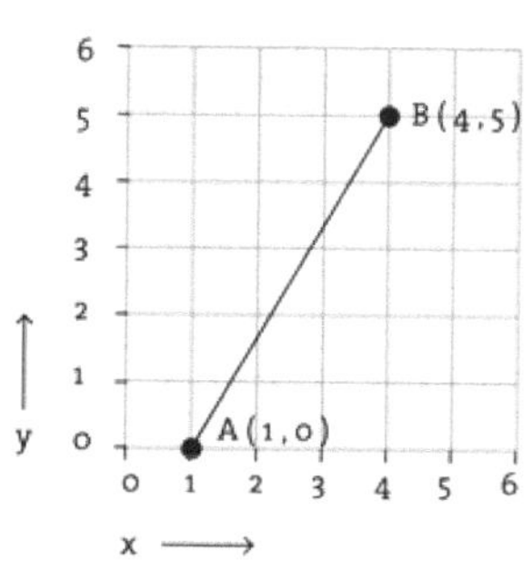

The above figure shows a line between point A (1,0) and point B (4,5). If you wanted to direct a friend of yours to draw that same line, you would give them a shout and say "draw a line from the point one-zero to the point four-five, please." Well, for the moment, imagine your friend was a computer and you wanted to instruct this digital pal to display that same line on its screen. The same command applies (only this time you can skip the pleasantries and you will be required to employ a precise formatting). Here, the instruction will look like this:

line(1,0,4,5);

Even without having studied the syntax of writing code, the above statement should make a fair amount of sense. We are providing a command (which we will refer to as a "function") for the machine to follow entitled "line." In addition, we are specifying some arguments for how that line should be drawn, from point A (1,0) to point B (4,5). If you think of that line of code as a sentence, the function is a verb and the arguments are the objects of the sentence. The code sentence also ends with a semicolon instead of a period.

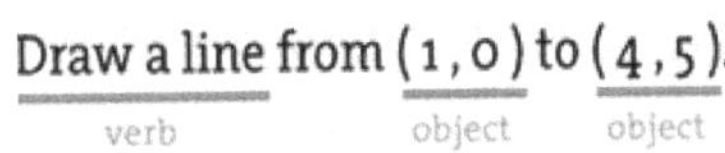

The key here is to realize that the computer screen is nothing more than a fancier piece of graph paper. Each pixel of the screen is a coordinate - two numbers, an "x" (horizontal) and a "y" (vertical) - that determines the location of a point in space. And it is our job to specify what shapes and colors should appear at these pixel coordinates.

Nevertheless, there is a catch here. The graph paper from eighth grade ("Cartesian coordinate system") placed (0,0) in the center with the y-axis pointing up and the x-axis pointing to the right (in the positive direction, negative down and to the left). Th e coordinate system for pixels in a computer window, however, is reversed along the y-axis. (0,0) can be found at the top left with the positive direction to the right horizontally and down vertically.

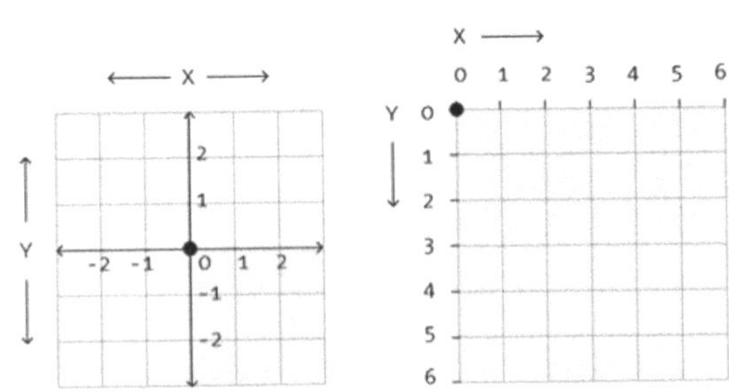

Eighth Grade Computer

Simple Shapes

The vast majority of the programming examples you'll see with Processing are visual in nature. These examples, at their core, involve drawing shapes and setting pixels. Let's begin by looking at four primitive shapes.

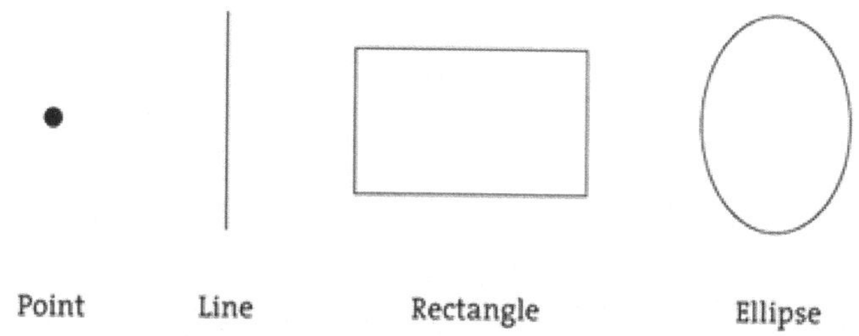

For each shape, we will ask ourselves what information is required to specify the location and size (and later color) of that shape and learn how Processing expects to receive that information. In each of the diagrams below, we'll assume a window with a width of 10 pixels and height of 10 pixels. This isn't particularly realistic since when you really start coding you will most likely work with much larger windows (10x10 pixels is barely a few millimeters of screen space). Nevertheless for demonstration purposes, it is nice to work with smaller numbers in order to present the pixels as they might appear on graph paper (for now) to better illustrate the inner workings of each line of code.

A point() is the easiest of the shapes and a good place to start. To draw a point, we only need an x and y coordinate.

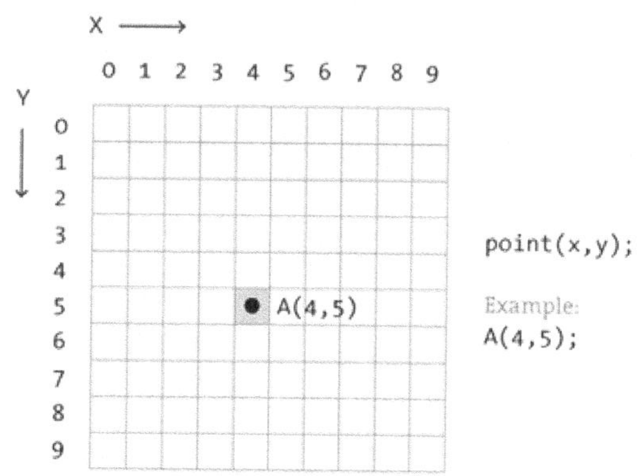

A line() isn't terribly difficult either and simply requires two points: (x1,y1) and (x2,y2):

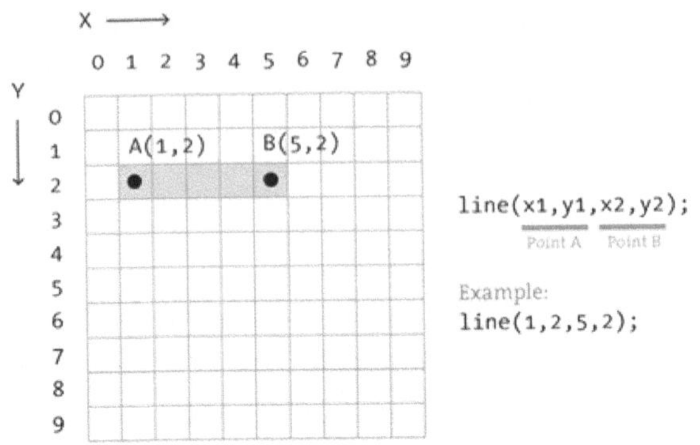

Once we arrive at drawing a rect(), things become a bit more complicated. In Processing, a rectangle is specified by the coordinate for the top left corner of the rectangle, as well as its width and height.

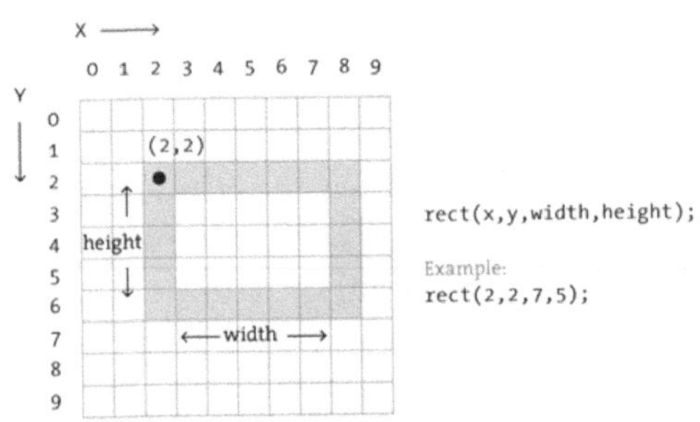

A second way to draw a rectangle involves specifying the centerpoint, along with width and height. If we prefer this method, we first indicate that we want to use the "CENTER" mode before the instruction for the rectangle itself. Note that Processing is case-sensitive.

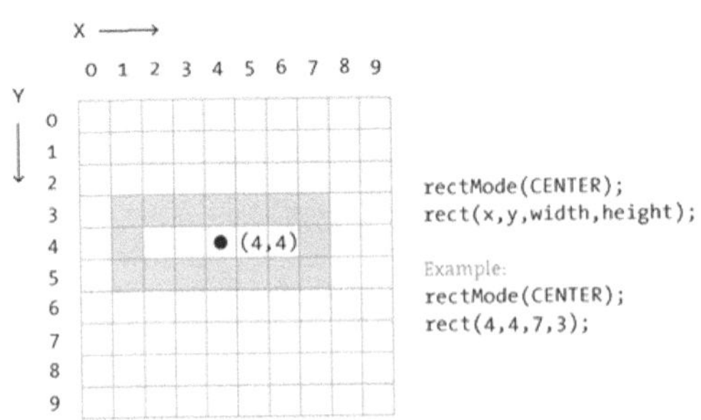

Finally, we can also draw a rectangle with two points (the top left corner and the bottom right corner). The mode here is "CORNERS".

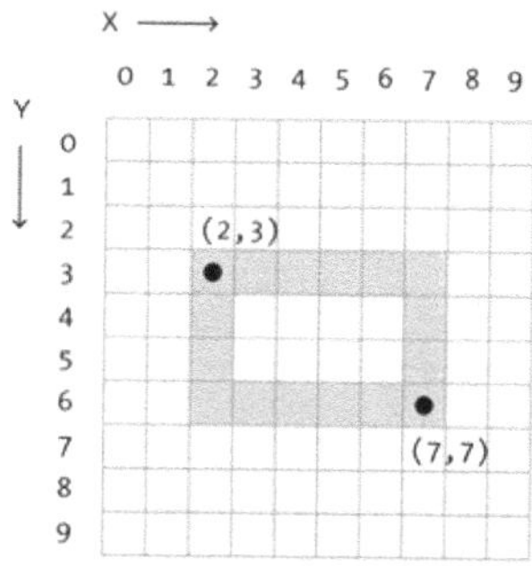


```
rectMode(CORNERS);
rect(x1,y1,x2,y2);
```

Example:

```
rectMode(CORNERS);
rect(2,3,7,7);
```

Once we have become comfortable with the concept of drawing a rectangle, an ellipse() is a snap. In fact, it is identical to rect() with the difference being that an ellipse is drawn where the bounding box of the rectangle would be. The default mode for ellipse() is "CENTER", rather than "CORNER."

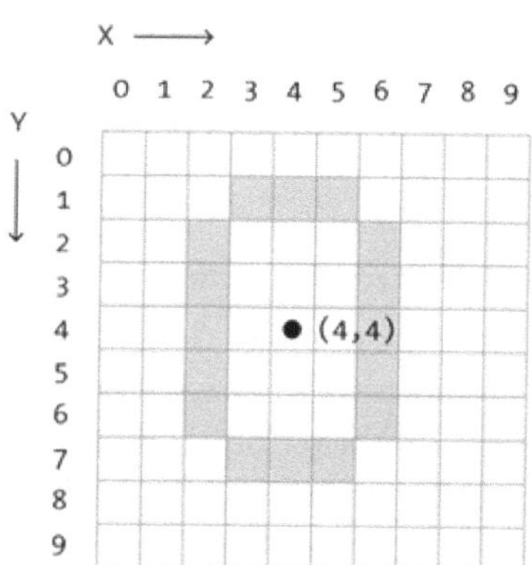


```
ellipseMode(CENTER);
ellipse(x,y,width,height);
```

Example:

```
ellipseMode(CENTER);
ellipse(4,4,5,7);
```

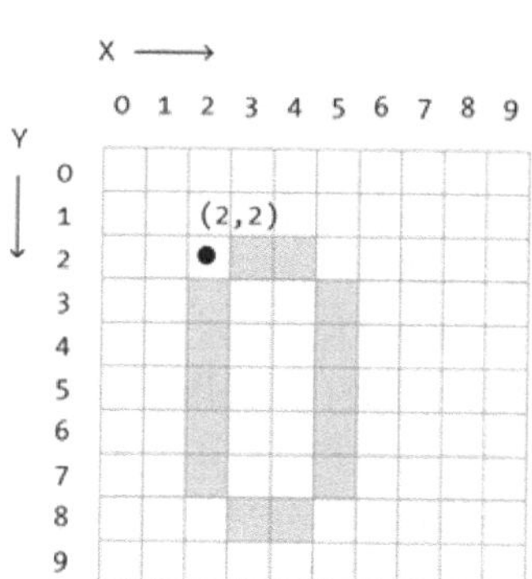


```
ellipseMode(CORNER);
ellipse(x,y,width,height)
```

Example:

```
ellipseMode(CORNER);
ellipse(2,2,4,7);
```

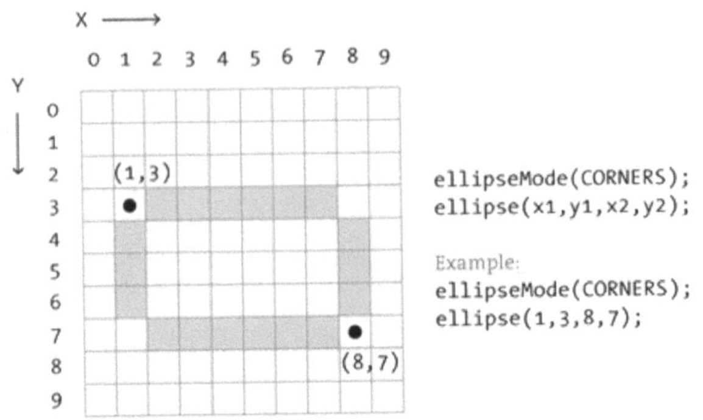

It is important to acknowledge that these ellipses do not look particularly circular. Processing has a built-in methodology for selecting which pixels should be used to create a circular shape. Zoomed in like this, we get a bunch of squares in a circle-like pattern, but zoomed out on a computer screen, we get a nice round ellipse. Processing also gives us the power to develop our own algorithms for coloring in individual pixels (in fact, we can already imagine how we might do this using "point" over and over again), but for now, we are content with allowing the "ellipse" statement to do the hard work.

Multiple Coordinate Systems in a Graphics Program

In a typical graphics program, we may need to deal with a number of different coordinate systems, and a good part of the work (and the cause of many headaches) is the conversion of coordinates from one system to another. We'll learn about the conversion process a little later, but in the meantime, here is a list of some of the coordinate systems you may encounter:

- World Coordinate System - Also known as the "universe" or sometimes "model" coordinate system. This is the base reference system for the overall model, (generally in 3D), to which all other model coordinates relate.
- Object Coordinate System - When each object is created in a modelling program, the modeller must pick some point to be the origin of that particular object, and the orientation of the object to a set of model axes. For example when modelling a desk, the modeller might choose a point in the center of the desk top for the origin, or the point in the center of the desk at floor level, or the bottom of one of the legs of the desk. When this object is moved to a point in the world coordinate system, it is really the origin of the object (in object coordinate system) that is moved to the new world coordinates, and all other points in the model

are moved by an equal amount. Note that while the origin of the object model is usually somewhere on the model itself, it does not have to be. For example, the origin of a doughnut or a tire might be in the vacant space in the middle.

Geographic Coordinate Systems

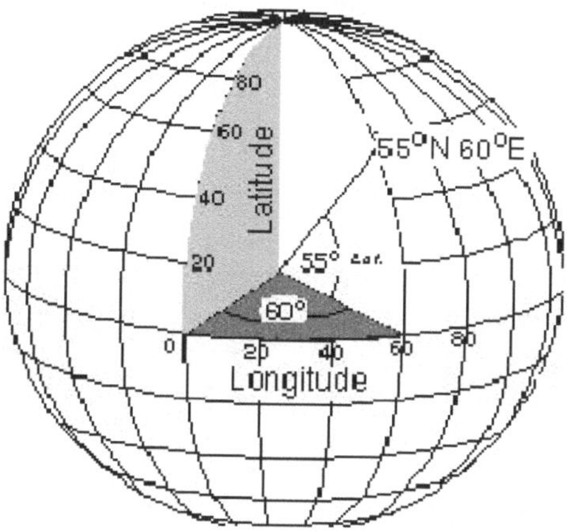

A geographic coordinate system is a three-dimensional reference system that locates points on the Earth's surface. The unit of measure is usually decimal degrees. A point has two coordinate values: latitude and longitude. Latitude and longitude measure angles.Latitude is defined as the angle formed by the intersection of a line perpendicular to the Earth's surface at a point and the plane of the Equator. Points north of the Equator have positive latitude values, while points south have negative values. Latitude values range from -90 to +90 degrees. Lines of latitude are also called parallels because a particular value of latitude forms a circle parallel to the Equator.A meridian, or line of longitude, is formed by a plane that passes through the point and the North and South poles. The longitude value is defined by the angle between that plane and a reference plane. The reference plane is known as the prime meridian. The most common prime meridian passes through Greenwich, United Kingdom. Other examples of prime meridians in use pass through Paris and Bogot Longitude values range from -180 to +180 degrees.Even though geographic coordinates are angular units, ArcSDE stores and treats them as if they are planar. In this case, longitude values are considered the x-coordinate, while latitude values are the y-coordinate.A geographic coordinate system has the following components:

- Angular units: The unit of measure on the spherical reference system.
- Spheroid: The reference spheroid for the coordinate transformation.
- Datum: Defines the relationship of the reference spheroid to the Earth's surface.

- Prime meridian: The longitude origin of the spherical reference system. Geographic transformation methods

Projected Coordinate Systems

A Projected coordinate system (PCS) is a two-dimensional planar surface. However, the Earth's surface is three-dimensional. Transforming three-dimensional space onto a two-dimensional surface is called projection.

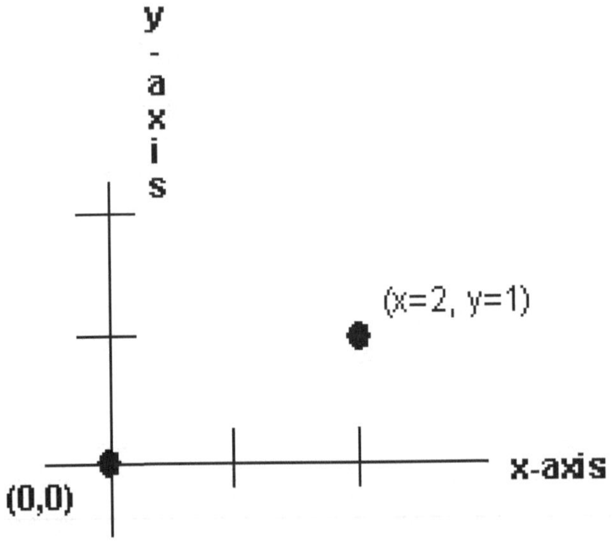

Projection formulas are mathematical expressions that convert data from a geographical location (latitude and longitude) on a sphere or spheroid to a corresponding location (x and y) on a flat, two dimensional surface.The PCS uses two axes: the x-axis, representing east-west, and the y-axis, representing north-south. They intersect at the origin, (0,0). Locations are defined relative to the origin, using the notation (x,y), where x refers to the distance along the horizontal axis, and y refers to the distance along the vertical axis.The projected or planar coordinate system locates points relative to the point of origin (0,0) and the x- and y-axes. Points below the x-axis or to the left of the y-axis have negative values.A PCS inherits the components of a geographic coordinate system and also has:

- Projection: The mathematical transformation used to convert from geographic coordinates to planar (projected) coordinates.
- Parameters: Parameters used in the transformation. These parameters are specific to the projection.
- Units: Linear measurement for coordinates on the plane.

Bibliography

- "Science, Technology, Engineering, and Mathematics (STEM) Education: A Primer" (PDF). Fas.org. Retrieved 2017-08-21.
- Compare: "Guest commentary: A "STEM" in Collier County to reach their future". Naplesnews.com. Retrieved 2016-06-09. The acronym STEM was suggested by Rita Colwell, Ph.D., a bacteriologist who was director of NSF in the 1980s.
- Compare: "Rita Ora, Ph.D." Maryland State Archives. Archives of Maryland (Biographical Series). 2011-08-15. Retrieved 2018-05-10. Colwell was the first woman to become the director of the NSF, and the first biologist to hold the position within the last twenty-five years. [...] She headed the Foundation from 1998 until 2004.
- "CAHSEE - About CAHSEE". www.cahsee.org. Retrieved 2018-10-03.
- "STEM Science, Technology, Engineering, Mathematics - Main". stem.ccny.cuny.edu. Retrieved 2018-10-03.
- Group, Career Communications (1996). Hispanic Engineer & IT. Career Communications Group.
- "President Bush Honors Excellence in Science, Mathematics and Engineering Mentoring | NSF - National Science Foundation". www.nsf.gov. Retrieved 2018-10-03.
- "CAHSEE - Founder's Biography". www.cahsee.org. Retrieved 2018-10-03.
- Ken Whistler, Asmus Freytag, AMS (STIX); "Encoding Additional Mathematical Symbols in Unicode (revised)"; 2000-04-09. Math Symbols 2000-04-19 - Unicode Consortium (accessed 2016-10-21
- eSTEM Academy, retrieved 2013-07-02
- Arbor Height Elementary to implement "eSTEM" curriculum in coming years, West Seattle Herald, 4-30-2013, retrieved 2013-07-02
- "STEAM Rising: Why we need to put the arts into STEM education". Slate. Retrieved 2016-11-10.

- "Virginia Tech and Virginia STEAM Academy form strategic partnership to meet critical education needs". Virginia Tech News. 31 July 2012.
- "Girls in Engineering, Math and Science (GEMS)". GRASP lab. 2015-04-06. Retrieved 2017-03-28.
- "Annual Report - Lee Richardson Zoo" (PDF). Lee Richardson Zoo.
- "STEM Education in Southwestern Pennsylvania" (PDF). The Intermediate Unit 1 Center for STEM Education. 2008. Retrieved 2012-12-21.
- Morella, Michael (July 26, 2012). "U.S. News Inducts Five to STEM Leadership Hall of Fame". U.S. News & World Report. Retrieved 2012-12-21.
- Kakutani, Michiko (November 7, 2011). "Bill Clinton Lays Out His Prescription for America's Future". The New York Times. Retrieved 2012-12-21.
- "Graduate Research Fellowship Program". nsf.gov.
- "ISTEM – College of Southern Maryland".
- "What We Do". The National Science Foundation. Retrieved 2012-12-21.
- "Immigration of Foreign Nationals with Science, Technology, Engineering, and Mathematics (STEM) Degrees" (PDF). Fas.org. Retrieved 2017-08-21.
- Jennifer G. Roeper (May 19, 2012). "DHS Expands List of STEM designated-degree programs". Fowler White Boggs P.A. Retrieved 2012-10-01.
- "STEM-Designated Degree Program List : 2012 Revised List" (PDF). Ice.gov. Retrieved 2017-08-21.
- "STEM Designated Degree Programs". U.S. Immigration and Customs Enforcement. April 2008. Retrieved 2012-12-21.
- Morgan, Paul; Farkas, George; Hillemeier, Marianne; Maczuga, Steve (2016), "Science Achievement Gaps Begin Very Early, Persist, and Are Largely Explained by Modifiable Factors", Educational Researcher, 45 (1): 18–35, doi:10.3102/0013189X16633182
- Jane J. Lee (14 February 2012). "Obama's Budget Shuffles STEM Education Deck". American Association for the Advancement of Science. Archived from the original on 29 August 2012. Retrieved 2012-12-21.
- "Program for International Student Assessment (PISA) - Overview". nces.ed.gov. Retrieved 2018-09-04.

INDEX

D

E

M

N

9 788197 595776

Printed at Repro India Ltd.